Formwork and Concrete Practice

Peter Brett

Hammersmith and West London College

Heinemann Professional Publishing

Heinemann Professional Publishing Ltd
22 Bedford Square, London WC1B 3HH

LONDON MELBOURNE AUCKLAND

First published 1988

British Library Cataloguing in Publication Data
Brett, Peter, *1950–*
Formwork and concrete practice.
1. Concrete and construction – Formwork
I. Title
624.1′834 TA683.44

ISBN 0 434 90177 6

Photoset by Deltatype, Ellesmere Port, Cheshire
Printed in Great Britain by
Richard Clay Ltd, Chichester, Sussex

Contents

Preface

Formwork and Concrete Practice covers the course requirements for the City and Guilds of London Institute (CGLI) 582 Formwork; CGLI 607 Formwork for Concrete Construction; CGLI 585 Advanced Craft Formwork option for carpenters and joiners; and CGLI 580 Concrete Practice. In addition this work will be found very useful by those studying on CGLI concrete technology and construction courses, as well as the formwork and concrete elements of building technician courses and overseas schemes.

The presentation of this book is designed to explain clearly and simply to students the often complex allied subjects of formwork and concrete practice. Each topic is covered from basic principles, thus making the book suitable for readers who have no previous knowledge of the subject as well as for those who wish to extend their understanding.

Although primarily intended for students, this book will also be of use to construction personnel at all levels, either as a source of information or to revise and update their existing knowledge.

Each of the chapters ends with a series of multiple choice or short answer self-assessment questions. By attempting these questions, readers will be evaluating their understanding of the relevant chapter, checking their progress through the course, and in addition undertaking valuable practice in answering examination, assignment and other assessment questions.

At the end of the book there is a guide to examination and study techniques.

Part One

Formwork

Chapter 1

Requirements and materials for formwork

Requirements of formwork

Formwork (sometimes referred to as *shuttering*) can be defined as a structure, usually temporary (but in certain circumstances partly or wholly permanent), which is designed to contain fresh fluid concrete; form it into the required shape and dimensions; and support it until it cures sufficiently to become self-supporting. The term 'formwork' includes the actual material in contact with the concrete, known as the *form face*, and all the necessary associated supporting structure.

The supporting structure is also referred to as *falsework*, more especially on large civil engineering contracts. A definition of falsework is any temporary structure used to support a permanent structure during its erection and until it becomes self-supporting. As such falsework is included in the definition of formwork, given above.

In order to successfully carry out its function, formwork must achieve a balance of the following essential requirements:

Containment Formwork must be capable of shaping and supporting the fluid concrete until it cures.

Strength Formwork must be capable of safely withstanding without distortion or danger the dead weight of the fluid concrete placed on it, the imposed loading of the construction workers and equipment used in placing the concrete in position, and any environmental loadings.

Resistance to leakage All joints in the form face must be either close fitting or covered with form tape to make them grout tight. If grout leakage occurs the concrete will be weak at that point. Leakage causes honeycombing of the surface, and may produce projecting fins (snots) which later have to be removed and touched up.

Accuracy Formwork must be accurately set out so that the resulting concrete product is in the right place and is of the correct shape and dimensions. The degree of accuracy concerned will be consistent with the item being cast; for example, foundation formwork is not normally required to have such a high degree of accuracy as formwork for superstructures.

Ease of handling Form panels and units should be designed so that their maximum size does not exceed that which can be easily handled by hand or mechanical means. In addition all formwork must also be designed and constructed to include facilities for adjustment, levelling, easing and striking without damage to the formwork or concrete.

Finish and reuse potential The form face material must be selected to be capable of consistently imparting the desired concrete finish (smooth, textured, featured or exposed aggregate etc.) whilst at the same time achieving the required number of reuses.

Access for concrete Any formwork arrangement must provide access for the placing of the concrete. The extent of this provision will be dependent on the job in hand, the amount of concrete being poured or placed at any one time, and the ease of carrying out the concreting operations.

Economy All formwork is very expensive. On

average about 35 per cent of the total cost of any finished concrete unit or element can be attributed to its formwork; of this just over 40 per cent is for materials and nearly 60 per cent is for labour. The formwork designer must therefore not only consider the maximum number of times that any form can be reused, but also produce a design that will minimize the time taken for erection and striking.

Materials for formwork

There is available a wide variety of materials which can be used for formwork construction. These vary from the most common timber and steel to plastic, rubber and in certain circumstances even concrete. The actual selection of the material to be used for a particular situation will be based on consideration of a number of technical, practical and economic factors.

Thus when selecting materials for formwork the following main points must be considered:

1 Strength of materials
2 Economic use of materials
3 Ease of handling, working and erection
4 Ability to form the desired shape
5 Facilities for adjustment, levelling, easing and striking
6 Quality of finish required.

Timber

Softwood is the most commonly used material for formwork. The reasons for this are its:

Availability Softwood is readily available from a wide number of outlets.
Economy Softwood is cheap relative to alternative materials.
Structural properties Softwood has adequate strength properties coupled with low weight.
Ease of working and handling Softwood is very easily worked using basic tools and equipment. Complicated, intricate and curved formwork shapes can be built up as required.
Insulation properties In cold weather softwood helps to retain the heat in curing concrete.

Softwood grades

Most of the softwood used for formwork is imported and is graded for quality at its source of origin by the exporting sawmill. Quality varies from country to country but in all cases the higher grades command the higher prices. Most softwoods are graded using a defects system which specifies the maximum amount or size of a defect permissible for each grade of timber.

A simplified definition of softwood gradings used by the main exporting countries is as follows.

North European countries
Sweden, Finland and other European countries have a unified grading system that describes six basic qualities of redwood and whitewood, numbered I, II, III, IV, V and VI (firsts, seconds, thirds, fourths, fifths and sixths). Qualities I, II, III and IV are rarely available separately; the common practice is to group them together and sell them as one grade called *unsorted* (US), which is classified as being suitable for joinery. The V quality is available separately and is classified for use in general construction and carcassing work; it is thus suitable for formwork construction. The VI quality is classified as a utility grade and is used mainly for packaging; although more economical than V, it is unlikely to be suitable for most formwork applications.

Russia
This grading system for redwood and whitewood is similar to that of the northern European countries except that only five basic qualities are used. I, II and III are exported as unsorted; this is approximately the same as the Swedish/Finnish unsorted. The Russian grades IV and V are similar or slightly better than the Swedish/Finnish V and VI. Thus both the two lower grades of Russian softwood are normally suitable for formwork construction.

Canada and USA
Their timber is often marketed in mixed species groups, such as hem/fir (a mix of western hemlock and amabilis fir) or spruce/pine/fir. The main

grades available are: selected merchantable and no. 1 merchantable, which are similar to the European unsorted qualities; no. 2 merchantable, which is a general carcassing quality suitable for formwork; and no. 3 common, which is a utility grade mainly for packaging.

Stress grading

As the strength properties of timber vary widely between species and even between different pieces cut from the same tree, wherever possible timber used for formwork should be stress graded to ensure its suitability. Stress or strength grading of timber classifies it for structural purposes and thus provides the formwork structural designer with a material of a known minimum strength.

Stress grades make allowance for the inevitable inclusion in any piece of timber of strength-reducing defects (knots, fast rate of growth, sloping grain, distortions etc.). BS 4978:1973 *Timber Grades for Structural Use* defines the following visual or mechanical grades of timber:

Visually graded
General structural (GS)
Special structural (SS)
Mechanically graded
Machine general structural (MGS)
Machine special structural (MSS)

In addition numbered mechanical grades of M50 and M75 are available. Therefore there is a range of six commercially available stress grades, which in descending strength order are:

M75, MSS/SS, M50, MGS/GS

Each piece of stress graded timber must be marked with its stress grade. In the case of machine graded timbers this may be by splashes of colour dye along the length of the piece. The standard recommended colour coding for this is:

M75	red
MSS	purple
M50	blue
MGS	green

Stress graded timber imported from Canada is stress graded under the National Lumber Grades Authority (NLGA) rules. This defines the following four grades of structural joists and planks:

Select structural (sels)
No. 1 structural
No. 2 structural
No. 3 structural

Strength class

BS 5268:part 2:1984 *Structural Use of Timber* uses the previous strength grades and further classifies them by grade and species into nine strength classes. Strength classes 1 to 5 cover the most common softwoods which are suitable for formwork construction. These are shown in Table 1.

Purchasing of timber

The actual quality of the timber purchased will depend on the nature of the work in hand, but timber will normally be selected from one of the following three categories:

Minimum specification: strength class 3 (SC3) This is the minimum standard of timber acceptable for structural use in formwork and falsework.
Standard specification: strength class 4 (SC4) This is the standard of timber recommended for general formwork use and is accepted as being good practice.
High specification: strength class 5 (SC5) This is recommended for use where a higher grade of timber is required for heavily loaded members.

Stress graded timber is available as either sawn or processed (regularized or PAR). The basic range of sawn stress graded sizes commonly available is shown in Figure 1.

When buying timber for formwork it is common practice to obtain either regularized or planed all round (PAR).

Regularized timber is that which is brought to a uniform width by machining on one or both edges. A reduction in width of 3 mm is allowed for timbers up to 150 mm, and a reduction of 5 mm over this width. For example, a 75 mm × 200 mm ledger may be regularized to 75 mm × 195 mm.

Table 1

Species	*Strength class (SC) and grade* 1	2	3	4	5
Imported softwood					
Parana pine			GS	SS	
Redwood			GS/M50	SS	M75
Whitewood			GS/M50	SS	M75
Western red cedar	GS	SS			
Douglas fir	No. 3		No. 1/No. 2 GS	Sels/SS	
Hem/fir	No. 3		No. 1/No. 2 GS/M50	Sels/SS	M75
Spruce/pine/fir	No. 3		No. 1/No. 2 GS/M50	Sels SS/M75	
British grown softwood					
Douglas fir		GS	M50/SS		M75
Larch			GS	SS	
Scots pine			GS/M50	SS	M75
Spruce	GS	M50/SS	M/75		

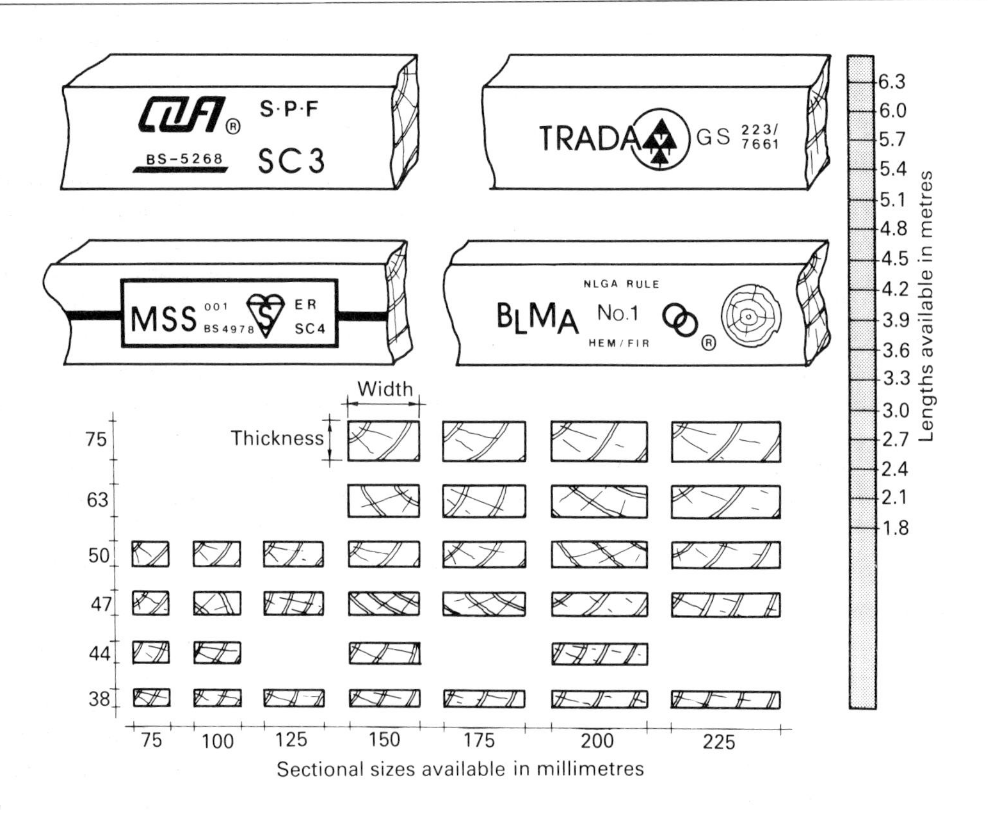

Figure 1 *Typical stress grade marks and common sizes*

PAR timber is preferred for formwork not only because of its uniform cross-section, but also because it is better to handle and easier to clean off after any concrete seepage or spillage. Canadian PAR is often supplied with its four arrises eased or pencil rounded. The normal reductions from the basic sawn size, for planing two opposite faces or edges, is 3 mm for dimensions up to 100 mm; 5 mm for dimensions between 100 mm and 150 mm; and 6 mm for dimensions over 150 mm. For example, a 75 mm × 200 mm may become 72 mm × 194 mm as PAR.

It is recognized that the moisture content of timber will have a bearing on its actual sectional size. The basic sizes refer to timber with a moisture content of 20 per cent. A tolerance on the width of ±1 mm is permissible for regularized timber, and of ±0.5 mm is permissible for PAR timber.

Standard lengths of softwood start at 1.8 m and rise in 300 mm increments. Timber from European sources is rarely available over about 6 m in length. Lengths of up to about 12 m are sometimes available from Canadian sources.

Use of timber on site

The use of stress graded timber on site must be closely controlled. There is a danger in having several grades on site in that a low grade may be unwittingly used in a highly stressed situation. Therefore all timber, whether new or used, must be readily identifiable as complying with the appropriate specification. This can be achieved by using the following good practice points:

1 Order only one specification of timber for a given site.
2 Order new timber for each site; or
3 Treat timber as an item of plant. It should then be clearly marked with its grade, de-nailed, cleaned and stored after use and regularly inspected for defects.
4 Where timber is resawn it should be borne in mind that cross-cutting does not affect its grade whereas rip sawing does. Therefore all timber that has been resawn to a smaller section must be regraded before use.
5 Likewise, stress graded timber that has lost its original markings must again be regraded before use.

The use of solid timber in formwork is almost exclusively restricted to the support and framing members; it is rarely used as the actual form face for soffits and sheathing. This is for reasons of both quality of finish and economics. The exception to this is where decorative board-marked feature finishes are required. In these cases the formwork is lined with either prepared or sawn softwood boards. A sand blasting treatment or a prior soaking in water will enhance the natural grain pattern, which is transferred to the finished concrete.

Where timber boarding is used as the form face its reuse potential (the number of times it can be used for casting whilst still producing an acceptable finish) will vary between one and ten uses. This depends on the quality of finish required and the care taken in striking and handling.

Plywood

Plywood is by far the most common sheathing and soffit material used for formwork. The two main types of plywood in common use for formwork are Douglas fir and birch. The grain pattern of Douglas fir plywood will be transferred to the concrete face and may be used as a decorative feature. A much smoother finish to the concrete is achieved where birch plywood is used. In both cases the adhesives used in the manufacture of plywood for formwork must be suitable for the wet conditions to which it will be subjected; therefore the plywood should be of weather and boil proof (WBP) quality.

Plywood grades

Plywood is graded according to the appearance of its outer face veneers, each face of a sheet being assessed independently. Grading rules vary depending on the origin of the plywood. Most European manufacturers base their veneer grades on recommendations issued by the International Organization for Standardization (ISO). A simplified definition of these face veneer grades is given in Table 2.

Table 2 *European plywood grading (ISO)*

Face grade	*Veneer quality*
E	Smooth cut veneers; practically defect free
I	Clear veneer; limited number of small sound knots and other minor defects are permitted
II	Repaired veneer quality; large number of sound knots permitted. Loose knots and other defects will be repaired with plugs or filler
III	Unlimited number of sound knots up to 50 mm in diameter are permitted. Loose knots, holes and shakes are permitted to a limited extent
IV	Almost all defects are permitted

In most circumstances it is unnecessary for both faces of the sheet to be of the same grade. Thus most manufacturers offer a wide range of face grade combinations, e.g. E/I, E/II, I/II, I/III, I/IV etc.; these refer to the front and back face veneers of the sheet respectively.

Canada and America both use different face veneer grades for their plywoods. However, plywoods from these sources are available in a number of sheet grades which are approximately equivalent to the European grades, as indicated in Table 3.

Table 3 *Comparison of plywood sheet grades*

Canada	*America*	*Europe (ISO)*
G2S (good two sides)	A/A	E/E
G1S (good one side)	A/C	E/IV
Select	B/C	I/IV
Sheathing	C/C (sheathing)	IV/IV

Purchasing plywood

The thickness of plywood used for formwork is almost universally 18/19 mm, although the range available is from 4.5 mm to 32 mm. The standard sheet size is either 1200 mm × 2400 mm or 1220 mm × 2440 mm depending on the source of supply. However, larger sheet sizes are normally available to order. Sheet grade E/IV, G1S or A/C are normally specified for use where a high quality smooth finish to the concrete is required. Grades I/IV, select or B/C can be used where minor surface irregularities are permitted in the concrete finish. Grades IV/IV, sheathing or C/C can be specified for one-off underground or covered work for economic reasons where the concrete finish is not of prime importance.

Surfaced plywoods can be specified for high quality repetitive work. These give a greater number of reuses and a smoother concrete finish; in addition they are resistant to the effects of moisture and abrasion. These surface treatments, which are factory applied, include barrier paints and varnishes; cellulose film impregnated with a phenolic resin; and glass fibre (GRP) overlay impregnated with a polyester or phenolic resin.

Use of plywood on site

On site any cut edges or holes must be sealed in order to prevent moisture absorption into the exposed layers, which would result in swelling and subsequent delamination. This can be achieved by the application of a barrier paint or other waterproofing agent.

Plywood must be handled with care. This is especially so when working with surfaced plywoods, as damaged faces are not easily repaired.

Plywood's reuse potential will vary widely depending on its surface treatment and handling. Typical reuse values assuming a reasonable amount of care in handling and striking are shown in Table 4.

Table 4

Plywood surface treatment	*Reuses (up to)*
Untreated	10
Barrier paint or varnish	20
Resin film finish	50
GRP overlay	100

Note A compatible release agent should always be used with both treated and untreated plywoods.

Various textured and profiled face plywoods are available which impart a featured surface to the finished concrete. Their reuse potential is very limited as they become progressively more difficult to clean and maintain after each casting. Plywood sheathing is particularly useful for producing curved forms. This is achieved by fixing the plywood to a rigid timber frame which has been constructed to form the required shape. The minimum radius of the curve is related to the thickness of the plywood. It is preferable to use two or three layers of thin plywood rather than one thick. Table 5 shows the minimum radii for a range of standard sheet thicknesses which are regarded as practical in formwork.

Table 5

Plywood thickness mm	*Minimum radii Across face grain mm*	*Minimum radii Along face grain mm*
6.5	700	1300
9	1600	2000
12	2500	3000
15	3600	4800
18	6000	7000

Note Certain individual plywood manufacturers may recommend tighter radii for specific products than those shown, although consideration must be made to the forces required to bend and restrain plywood below these radii.

Plywood is a laminated construction that has more layers parallel to the face grain than across it; thus it is stronger parallel to the face grain. The face grain should be placed at right angles to the supporting members to ensure maximum strength. Where this is impractical or would increase the cutting or cause additional waste, as in the case of beam sides and column forms, extra supports can be added (see Figure 2).

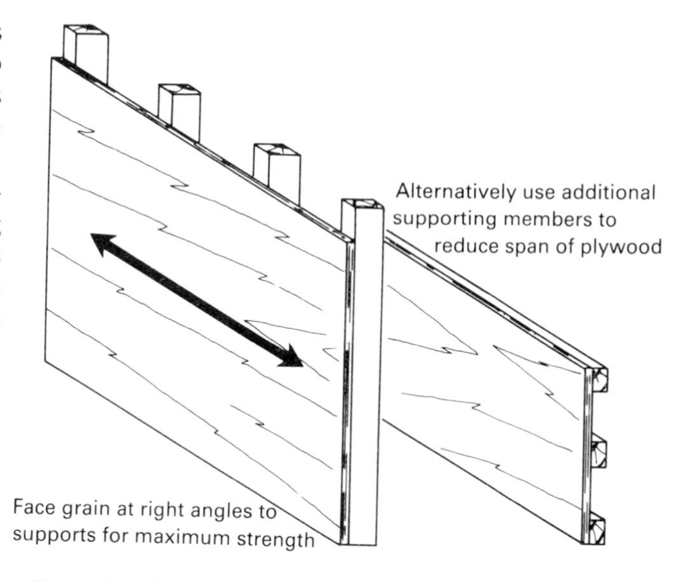

Figure 2 *Plywood span*

Other boards

Chipboard (particle board)

Chipboard is manufactured in thicknesses of 12 mm and 18 mm and supplied in sheets 1220 mm × 2440 mm with square edges. Also available are 600 mm × 2440 mm sheets with tongue and groove edges. Only exterior grade or special quality chipboard should be used for formwork, as the standard grades are not sufficially resistant to moisture absorption. Their use would lead to large amounts of moisture movement and a substantial loss in strength. Special film faced chipboards are also available for formwork sheathing and soffits. They have a high reuse potential provided they are properly treated and handled. The reuse potential of chipboard is less than that expected from an equivalent grade of plywood. This is due to the abrasive action of the concrete being poured causing a marked roughening of the surface.

Fibreboard

This is mainly used as a form face lining material and is either nailed or glued to a supporting backing of timber or plywood. It is particularly useful for forming curved surfaces, irregular features, circular or shaped columns etc. In addition owing to its fine grain-free nature it is useful for stop end construction where it is easily scribed around steel reinforcement etc., providing it is suitably stiffened with a supporting framework to resist concrete pressures.

The main type of fibreboard used for formwork

is hardboard. It is available in thicknesses of 3 mm and 6 mm and a wide range of standard sheet sizes. Two main qualities are used for formwork – standard hardboard and oil tempered hardboard. Oil tempered boards have a smoother finish, are less absorbent and thus produce a better concrete finish. These boards are also stronger and stiffer but more brittle than the standard boards. Therefore standard boards tend to be used for curved formwork as they can be bent to tighter radii.

The reuse potential of hardboard is fairly limited. Standard boards are considered suitable for one use only. Oil tempered boards, if properly treated, may be suitable for up to ten uses. The quality of the concrete finish will however rapidly deteriorate with each reuse.

Medium fibreboards and softboards are sometimes used as absorbent form liners. These impart a fine surface texture to the concrete finish and at the same time reduce the risk of blow holes. These boards are suitable for one use only.

Wood wool

Wood wool slabs are produced in thicknesses ranging from 25 mm to 100 mm. They are used as permanent formwork sheathing or soffit. In addition, permanent wood wool waffle and trough forms are available. The open texture of wood wool slabs provides an ideal key for both the cast concrete and any subsequent applied wall or ceiling finish. Wood wool slabs have the added advantage of their excellent thermal, acoustic and fire-resisting properties.

Metal

Both steel and aluminium are used for proprietary and special purpose-made formwork. The systems are constructed so that their component parts can be simply bolted or clipped to each other and to the supporting falsework. Where there is a high degree of repetitive work their use can be an economic alternative to traditional timber and plywood forms.

Proprietary metal forms

Proprietary metal formwork systems are available for hire from a number of specialist formwork suppliers. These will supply a wide range of standard equipment including the following:

1. Steel framed panels with either metal or plywood sheathing, which can be used for walls, columns, slabs, beams, pad bases etc.
2. Aluminium beams for use as ledgers and joists in slab construction and walings and soldiers in wall construction
3. Steel floor centres and strongbacks for slab and wall construction
4. Adjustable steel props, table forms and other support systems for slab construction
5. Column clamps, beam clamps, wall ties and an endless variety of other ironmongery and fittings designed to secure formwork.

Special purpose-made metal forms

Rather than being a standard stock item as are proprietary forms, these are generally exclusively manufactured for one specific contract. They tend to be used more for civil engineering contracts than for construction contracts. Typical applications of purpose-made forms include retaining walls, sea walls, slip forming, tunnelling etc.

This type of form will only be an economic consideration where large numbers of identical components are required, e.g. precast mould work for factory production; column forms for square, curved or circular work; and complicated shapes such as splayed or mushroom headed circular columns.

Use of metal forms

Metal formwork has a very high reuse potential. Provided it is correctly handled and thoroughly cleaned, oiled and maintained after each use, from 100 to 600 reuses can be expected. This will of course depend on the location of the concrete component and the standard of finish that is acceptable.

Where metal formwork has become damaged, distorted or otherwise defective, it must be taken out of service for reasons of safety as well of the accuracy and standard of finish to the concrete component. The repair of these forms is a highly specialized operation. No attempt should be made on site to straighten or reweld them. They should be either discarded or returned to the manufacturer for repair or replacement.

Advantages of metal formwork
The main advantages gained when using metal forms are:

1 High strength, in that they are less likely to fail under load
2 Robustness, which gives them the ability to withstand the occasional rough handling on site
3 Consistently smooth concrete finish
4 Consistent accuracy in the size of the component
5 Ease of erection and supervision
6 Economy for repetitive use
7 Availability for hire for specific periods, thus enabling costs to be accurately assessed and controlled.

Disadvantages of metal formwork
Sometimes metal forms have disadvantages over other formwork materials. The main disadvantages are:

1 Lack of adaptability, as metal forms are made to set sizes; any make-ups must normally be infilled with timber and plywood
2 Difficulty in fixing inserts, box-outs, fixing blocks etc.
3 High weight compared with other formwork materials, making cranage necessary for both erection and striking
4 Low insulation value, causing high heat losses during curing and thus increasing the likelihood of frost damage
5 Impermeable surface, which almost inevitably will cause blow holes (very small holes appearing in the concrete finish caused by air bubbles being trapped against the form face).

Plastic

The use of both sheet and formed plastics for formwork can be an economical consideration where high quality finishes coupled with repetitive use and complicated shapes are required.

The advantages of the material are low weight; dimensional stability; high potential reuse value; ability to form complex shapes; and ability to produce a high quality defect-free finish. The disadvantages are very high manufacturing costs in labour, material and supervision; susceptibility to impact damage and surface scratching; and high deflection under load unless adequately stiffened with ribs.

Glass fibre reinforced plastics (GRP) are widely used in formwork construction for:

1 Trough and waffle moulds for slab construction
2 Profiled form liners, mainly for wall construction, which can provide a variety of intricate moulded patterned finishes
3 Special purpose-made forms for complex shapes such as mushroom headed circular columns etc.

Foamed plastic (polystyrene) is used as a disposable formwork material for a variety of applications including making pockets for holding-down bolts, and forming voids, holes, apertures and other ornamental features. It has the advantage of being easily removed by poking out, burning out or being dissolved in petroleum spirit.

Use of plastic forms
The reuse potential of plastic forms is fairly high. In excess of 100 uses can normally be achieved, although (in common with other formwork materials) this will be dependent on correct handling and treatment during and after use.

As plastics are particularly susceptible to both impact damage and surface scratching, great care must be taken to avoid hitting or using an abrasive action on their surface.

After striking the plastic form face must be immediately cleaned to remove all traces of the cement dust and paste. This may be carried out using a special cleaning agent, clean soapy water, an oil soaked rag, or as directed by the manufacturer. Any hardened cement paste should be removed with a wooden rather than a metal scraper. The use of metal scrapers should be avoided at all costs as these will scratch and gouge out the form surface.

Any slight damage to plastic forms can often be repaired successfully on site using a resin compound. Major repairs should be avoided as these are considered to be a specialist job for the manufacturer.

Other formwork materials

A variety of other materials is used for formwork construction. Some of the following are often applied as a form face lining to stronger sheathing and soffit backings in order to form a smooth textured or patterned concrete finish, whilst others are used as formwork accessories to create voids, holes, stop ends etc.

Moulded rubber and corrugated cardboard can be used to impart three-dimensional designs to a concrete surface.
Plastic laminates, polythene sheeting and building paper can be used as waterproof liners, giving a glass-smooth finish to the concrete as well as facilitating formwork striking.
Cardboard cylinders can be used as one-off column forms and are available in a variety of lengths and diameters. They have the added advantage if required of being able to be left in position during the contract as a protection to the concrete surface.
Helically wound cardboard tubes can be used for forming holes in walls, slabs etc. On striking they are simply pulled from one end and will unwind with ease.
Inflatable rubber or plastic tube forms can be used to form holes through walls and slabs. In addition they are also useful as day joint or construction joint stop ends in wall form construction. When used as stop ends the tube is placed between the reinforcing bars; on inflation it will squeeze out between the steel and form a grout tight fit against battens which have been planted on the form face.

Combined formwork systems

It should be apparent from the previous information that most formwork systems are in fact made from a combination of different materials rather than a single one. Most proprietary formwork systems are a combination of metal and steel. A common combination is steel framed and plywood covered panels, which are simply bolted together on site and used for walls, beams, columns and slab soffits. Another is aluminium ledgers, joists and soldiers with timber inserts on to which plywood sheathing or decking is fixed for use as wall forms or slab soffits. Even traditional timber and plywood formwork uses steel clamps for columns, metal form ties for walls, and adjustable steel props for plumbing both walls and columns in addition to their use as support systems for decking.

Release agents

Release agents are materials which are applied to the form face to prevent concrete sticking to the form and also to facilitate striking. The correct use of release agents has a significant effect on the quality of the concrete finish and the reuse potential of the formwork.

A wide variety of release agents is available for use. Their selection will be dependent on the form face material and the quality of concrete finish required. Manufacturers' advice should be sought when choosing a release agent to ensure that it is compatible not only with the form face but also with any other formwork preparation such as waxes, surface retarders, barrier paints, fillers etc.

The main categories of release agent and surface coatings in general use can be classified as follows:

Neat oils are either vegetable or mineral oils. They are cheap and readily available. Unfortunately they encourage the formation of blow holes and therefore are not generally recommended for use on any quality work.
Neat oil with surfactant A surfactant is a surface wetting agent which allows the oil to spread

uniformly over the form face and thus minimizes the formation of blow holes. It has good penetration and is fairly resistant to most climatic conditions.
Mould cream emulsion is an emulsion of water in oil. It is suitable for use as a general release agent for most formwork surfaces, particularly timber and other absorbent materials. This emulsion minimizes the risk of blow holes on the surface of the concrete but is easily removed by even brief rain showers prior to casting.
Water soluble emulsion is an emulsion of oil in water which causes discoloration and retardation of the cement paste and is therefore not recommended for any visual concrete.
Chemical release agents are made from a small amount of chemical suspended in a petroleum distillate. When applied to the form face the petroleum carrier evaporates leaving a residue film on the surface. This causes a chemical reaction at the form face, thus minutely retarding the concrete surface. These agents are recommended for all high quality work, although as the thin film has no sealing properties it is recommended that prior wax sealing of the forms is carried out to permit maximum formwork reuse. Chemical release agents provide a non-slip finish, thus minimizing any risk of persons slipping on the formwork and the possibility of transfer of the release agent to the steel reinforcement.
Surface sealers and coatings These are barrier coats of paint, lacquer or other impermeable substances which are applied to new timber and plywood form faces before their first use. These protect the surface from abrasion and moisture absorption, thus extending their reuse potential. They are pretreatments and are not recommended for use without the subsequent application of a conventional release agent before each casting.
Waxes Again these are pretreatments, particularly suitable for sealing absorbent form linings prior to their first use. They are also particularly useful for filling small imperfections in the form face. For example, new steel form faces should always receive a wax pretreatment prior to use as it fills the tiny pits in the surfaces of sheet steel; removes rust cells from the surface; reduces mill scale blemishes; and generally gives the form face a far more suitable surface for the subsequent application of release agent.
Surface retarders These are chemical solutions that are applied to the form face. Although they are not release agents they do have the same effect as chemical release agents in that they retard the setting of the surface concrete. Surface retarders are in fact used to produce exposed aggregate finishes, the surface cement being brushed or washed away after the formwork has been struck.

Table 6 relates the main categories of release agents and surface coatings in general use to the various types of material used as form faces, together with their recommended method of application.

Use of release agents

Whatever the type of release agent being used, the manufacturer's instructions regarding storage, method of application and rate of coverage should be followed carefully. The best results are achieved with a fairly thin film of release agent. Excessive application results in a poor concrete finish, increases the risk of it getting on to the steel reinforcement, and at the same time creates a safety hazard owing to the risk of people slipping on horizontal and sloping form faces. Most release agents have a rate of coverage of between 6 and 8 m^2 per litre.

One application of release agent is sufficient for new non-absorbent form faces, although new absorbent ones may require a subsequent application after the first has been allowed to soak in. Alternatively absorbent form faces can be presealed. Subsequent coats of release agent will be required prior to each casting.

After coating with release agent the prepared formwork should be protected from dust and dirt which may otherwise stick to the formwork surface. In addition release agents are susceptible to removal or evaporation by rainwater, strong sunlight or drying winds. Ideally the concrete should be cast within 24 hours of applying the

Table 6

Type	*Method of application*			*Suitable form face*				
	Brush	Squeegee	Spray	Sawn timber	Prepared timber	Plywood	Metal	Plastic
Neat oil	Not recommended for high quality work							
Neat oil with surfactant	*		*	*	*	*	*	*
Mould cream emulsion	*	*		*	*	*		
Water soluble emulsion	Not recommended for high quality work							
Chemical release agent			*	*	*	*	*	*
Sealers and coatings	Best factory applied			*	*	*	*	*
Waxes	As directed by manufacturer			*	*	*	*	*
Surface retarders	*	*	*	*	*	*	*	*

release agent. Chemical release agents are normally considered preferable where longer waiting periods are anticipated.

In cases where the steel reinforcement becomes contaminated by release agents it is essential that it is completely removed before casting, otherwise the bond between the steel and concrete will be affected. When reinforcement is contaminated, a rag soaked in a water soluble degreaser can be rubbed along the affected bars, which are then sprayed with water to remove all traces of both the release agent and the degreaser.

Precautions in use of release agents

Release agents contain substances which can be harmful to health. Skin contact can lead to a defatting of the skin and may result in dermatitis. Some are poisonous if swallowed, whilst others can cause narcosis if their sprayed vapour is inhaled. These and other harmful effects can be avoided if the proper precautions are taken:

1. Always follow the manufacturer's instructions with regard to their use and storage.
2. Never water down or mix release agents with other materials or allow them to become contaminated.
3. Never store release agents in extremes of temperature; agents with a water content will freeze, and chemical agents are highly flammable.
4. Avoid contact with skin, eyes or clothing.
5. Avoid breathing in the fumes, especially when spraying.
6. Avoid contact with foodstuffs to prevent contamination.
7. Always wear protective clothing, gloves or barrier cream for the hands and a respirator when spraying.

8 Never smoke or use near a source of ignition.
9 Provide adequate ventilation when used internally.
10 Always wash hands thoroughly before eating and after work with either soap and warm water or an appropriate hand cleanser.
11 In cases of skin discomfort, accidental inhalation of vapour, swallowing or contact with the eyes, medical advice must be sought without delay.

Determining quantities

The quantities of material required for any operation are dependent on the number of items being cast, the minimum permissible striking time, and the maximum time allowed for the work.

Number of forms required

Example
To determine the number of column forms required to cast 400 columns in 40 working days at a temperature of 16°C:

number of columns to be cast each day

$$= \frac{\text{total number of columns}}{\text{number of days allowed}} = \frac{400}{50} = 8$$

Thus 8 columns per day are required. The minimum striking time at 16°C is 12 hours. Therefore forms cast one day may be struck the next.

From the above, sufficient formwork for two days is required – one day for striking and re-erection of formwork, and one day for casting/curing. Therefore 16 sets of column forms are required.

Table 7 is an extract from a typical stage programme to cast the column layout illustrated in Figure 3. This shows that the formwork which was erected on day 1 and cast on day 2 is to be struck and re-erected on day 3.

Type of plywood

The type of plywood used for the form face is dependent on the number of times the forms will be reused.

Example
To determine the type of plywood to be used for casting 400 columns using 16 sets of forms:

number of reuses

$$= \frac{\text{total number of columns}}{\text{number of forms made}} = \frac{400}{16} = 25$$

Therefore as each form will be reused 25 times, film faced plywood must be specified (see Table 4).

Table 7

Operation	*Day* *1*	*2*	*3*	*4*	*5*	*6*
Erect steel reinforcement	1	2	3	4	5	6
Erect formwork	1	2	3	4	5	6
Cast concrete		1	2	3	4	5
Strike formwork and prepare for reuse			1	2	3	4

Note: Kickers to columns will require casting several days before erecting the column forms

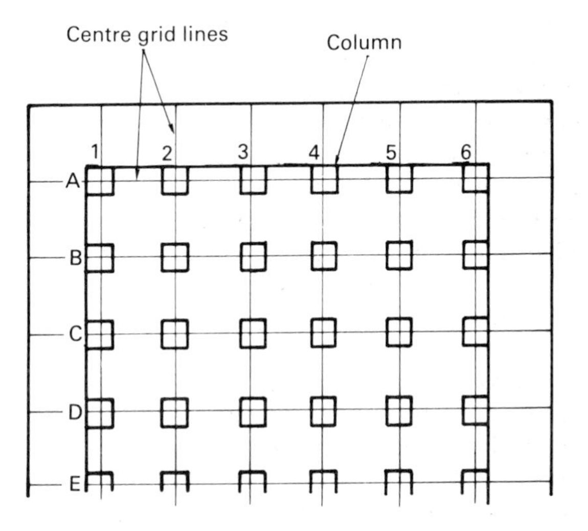

Figure 3 *Column grid layout*

Material lists

The quantities of material required for the column forms can be determined using the previous information and details of the formwork.

Example

To determine the materials required for the 250 mm square × 2400 mm high column form shown in Figure 4:

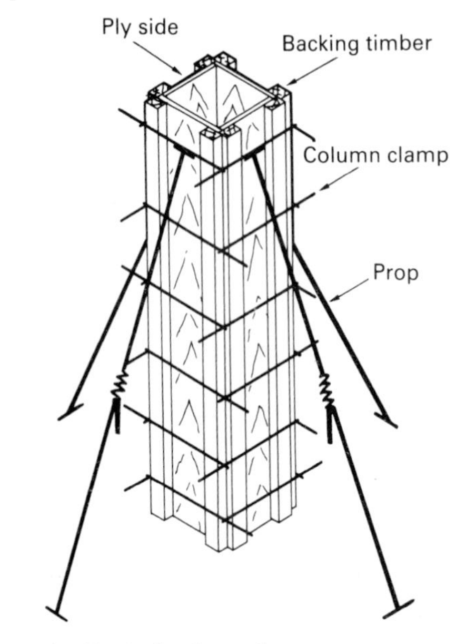

Figure 4 *Typical column form*

2 plywood sides at 18 mm × 250 mm × 2400 mm
2 plywood sides at 18 mm × 286 mm × 2400 mm

As the total width of the four column sides is less than 1200 mm, they can be cut from one sheet.

8 off softwood backing timbers 50 mm × 100 mm × 2400 mm
6 sets of column clamps
4 adjustable steel props
120 off 38 mm no. 8 countersunk steel screws (fixing plywood to backing timbers)
Quantity of 100 mm wire nails, say 1.5 kg (supporting column clamps and fixing prop heads)
Quantity of chemical release agent, say 0.325 litre (2.6 m^2 at 8 m^2 per litre)

As 16 sets of forms are required, the previous amounts must be multiplied by 16:

16 off 18 mm × 1200 mm × 2400 mm film faced plywood
128 off 50 mm × 100 mm × 2400 mm softwood
96 sets of column clamps
64 adjustable steel props
24 kg 100 mm wire nails (say one 25 kg box)
1920 off 38 mm no. 8 countersunk screws (say ten boxes of 200)
130 litres of chemical release agent (this quantity is sufficient to cast the 400 columns)

Self-assessment questions

Question	*Your answer*
1 State the grades of timber and plywood formwork suitable for a heavily loaded, high quality concrete finish.	
2 Describe a situation where the use of formed plastics for formwork may be justified.	
3 State *two* advantages and *two* disadvantages of using steel formwork in preference to timber formwork.	
4 Identify the following abbreviations: (a) M75 (b) SC5 (c) G1S (d) GRP.	
5 Name *three* formwork release agents, and for *each* state a suitable form face and its method of application.	
6 Briefly explain *five* general safety precautions to be observed during the storage and use of formwork release agents.	

7 If the reuse potential of a film faced plywood is given by its manufacturer as 40, then the minimum number of column forms required to cast 450 columns would be:
(a) 11
(b) 12
(c) 22
(d) 24.

a [ ]	b [ ]	c [ ]	d [ ]

8 Distinguish between formwork and falsework.

9 List *four* main requirements of formwork.

10 Explain *two* of the requirements given in the answer to question 9.

Chapter 2

Basic formwork design

Formwork must be designed to fulfil the requirements considered in Chapter 1, namely:

Containment
Strength
Resistance to leakage
Accuracy
Ease of handling
Finish and reuse potential
Access of concrete
Economy.

In determining an acceptable design solution, the formwork designer will seek to achieve a balance of these requirements in order to produce a formwork arrangement that satisfactorily contains, shapes and supports the fluid concrete, within any specified size tolerances, and with maximum economy of both labour and materials. At all times the safety of site personnel and the structural integrity of the formwork will be of prime importance.

Loads and pressures on formwork

The loads and pressures on formwork may be divided into four:

Dead loads consist of the self-weight of the formwork and the dead weight of the concrete and its steel reinforcement.
Imposed loads include the impact and surge loads of the concrete being placed and the live loads of the construction workers and concreting equipment, in addition to the later storage of materials such as bundles of reinforcement, stacks of timber and support systems etc., and forces from the permanent structure.
Environmental loads can be considered as wind loadings and accidental overloadings.
Hydrostatic pressure is caused by fluid concrete acting against the sides of vertical or steeply sloping formwork.

For convenience these are separated into wall type loadings, deck type loadings and hydrostatic pressure.

Note The loads and pressures on formwork are measured in newtons (N) and not kilograms (kg). Owing to the earth's gravitational pull, a mass of 1 kg will exert a force of approximately 9.81 N. This force will vary slightly from place to place on the earth's surface. For most practical purposes it is sufficiently accurate to take the weight of 1 kg to be 10 N; this simplifies calculations and at the same time errs on the safe side. Where fairly large numbers are concerned, kg can be converted to kN by dividing by 100. For example, 200 kg = 2 kN.

Walls and columns

The dead loads exerted on walls and columns (Figure 5) is the horizontal pressure of the fresh fluid concrete. This is in fact its hydrostatic pressure. Imposed loadings mainly occur from the surge or impact loading of the falling concrete, for which an allowance of 10 kN/m^2 is normally made. Environmental loads – the wind loads on vertical formwork – will vary according to the degree of exposure and the height of formwork. For any very exposed sites or where

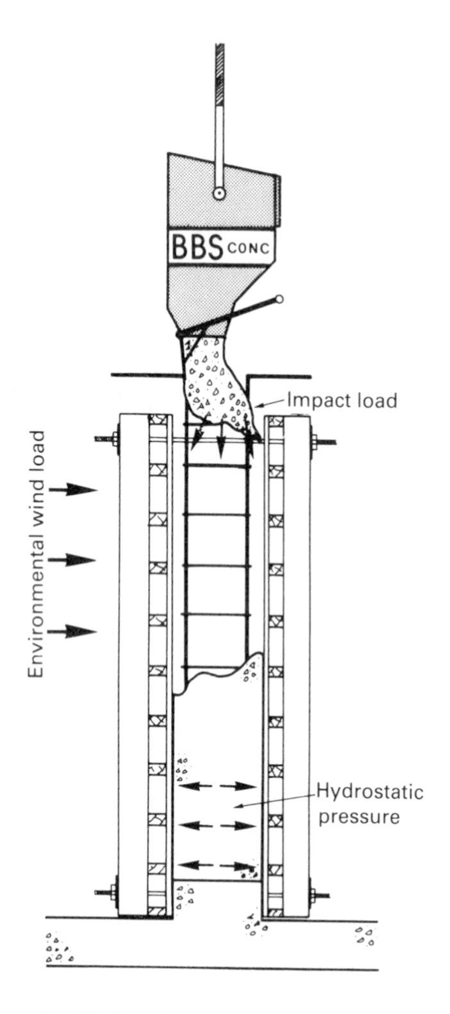

Figure 5 *Wall and column loadings*

major works are being undertaken, detailed calculations should be carried out to determine wind pressure. In practice the bracing members of wall and column formwork are sufficient to withstand the occasional accidental overload such as the touching of skips etc.; however, should this occur it is most important that a check for line and plumb is made.

Slabs and beams

These can be divided into vertical and horizontal loads.

Vertical loads

Dead vertical loads (Figure 6) on slabs and beams will vary according to the density of the concrete mix. For design purposes, assuming concrete with normal aggregates, a density of 2400 kg/m^3 is normally used. This has to be divided by 100 to obtain the pressure in kN/m^2 exerted on the formwork by a cubic metre of concrete. To determine the actual pressure exerted on the formwork it is necessary to take into account the slab thickness. This may be expressed as:

$$\underset{(\text{kN/m}^2)}{\text{pressure}} = \frac{\underset{(\text{kg/m}^3)}{\text{concrete density}} \times \underset{(\text{m})}{\text{slab thickness}}}{100 \text{ (converted to kN)}}$$

Example

The pressure exerted by a 150 mm thick concrete slab on the formwork is:

$$\text{pressure} = \frac{2400 \times 0.15}{100} = 3.6 \text{ kN/m}^2$$

The self-weight of the timber formwork is normally ignored. The imposed vertical load is the temporary load of construction workers and concreting equipment. This is taken to be anything between 1.5 and 3.5 kN/m^2 according to the circumstances in hand. In practice the higher figure is often used for decking as, in addition to the construction workers and concreting equipment, it also allows for the impact and heaping of concrete during placing and the storage of an amount of materials. The live loads on other structural members (e.g. joists, ledgers and props) is spread over a wide area, making its effect less concentrated. Therefore a value of 2 kN/m^2 is normally used for this purpose. Where space is restricted and it is intended to store large amounts of material on the slab, an allowance for the additional storage load should be made.

Example

The total unit design load on decking from a 150 mm thick slab would be:

$$\underset{(\text{kN/m}^2)}{\text{design load}} = \text{dead load} + \underset{(\text{kN/m}^2)}{\text{imposed load}}$$
$$= 3.6 + 3.5 = 7.1 \text{ kN/m}^2$$

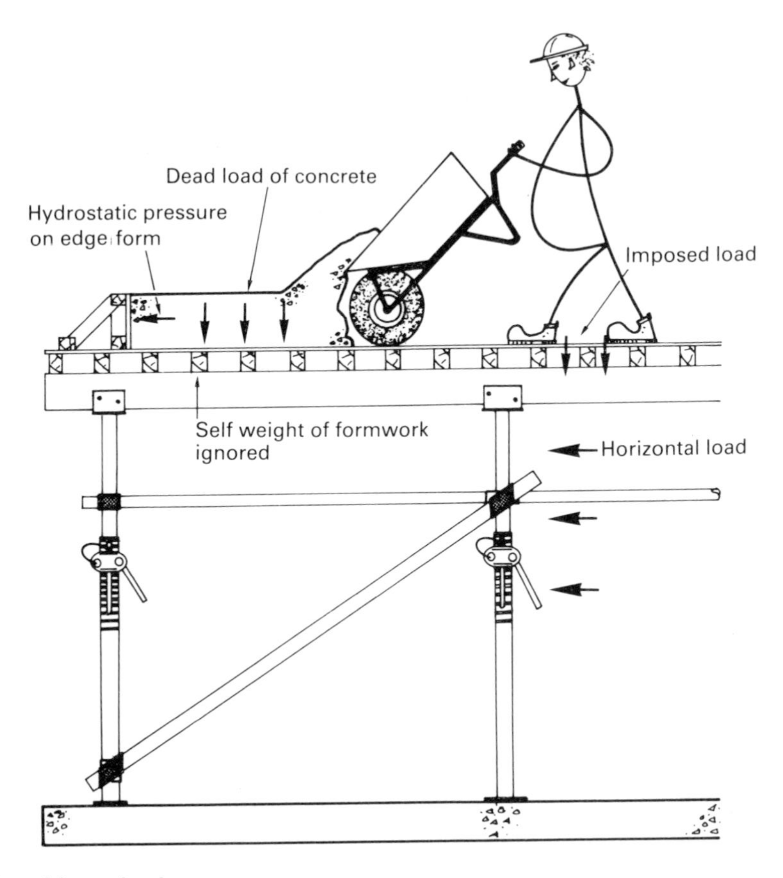

Figure 6 *Slab and beam loads*

Example
The total unit design load on joists, ledgers and props from a 150 mm thick slab would be:

$$\text{design load} = \text{dead load} + \text{imposed load}$$
$$(\text{kN/m}^2) \qquad (\text{kN/m}^2)$$
$$= 3.6 + 2 = 5.6\ \text{kN/m}^2$$

Horizontal loads
These result from pressure on slab and formwork, out of plumb support systems, and movement of plant and materials on the decking. Calculation of this loading is not normally required where the supporting structure is adequately laced and braced and less than 6 m in height.

Hydrostatic pressure
Freshly poured unset concrete acts as a liquid which exerts pressure (known as hydrostatic pressure) on all surfaces it touches. This pressure is caused naturally by the weight of the concrete which is under the influence of the earth's gravitational force. The amount of pressure is dependent on the height of the pour, and therefore is only significant in steeply sloping or vertical formwork.

The pressure in liquids increases with depth, and at any point it will always be in direct proportion to the height of liquid above. Thus the maximum hydrostatic pressure at any point within a form will be equal to the amount of liquid concrete above it. There will be maximum pressure at the bottom of the form, and the pressure will gradually decrease to a minimum at the top of the form.

The simple approach in determining the pressure at the *bottom* of a form is to multiply the concrete's density (2400 kg/m^3) by its height.

Example
The pressure at the bottom of the 1 m × 1 m × 4 m high column form in Figure 7 will be:

Pressure at bottom
= density of concrete × height of concrete
= 2400 kg/m^3 × 4 m = 9600 kg/m^2

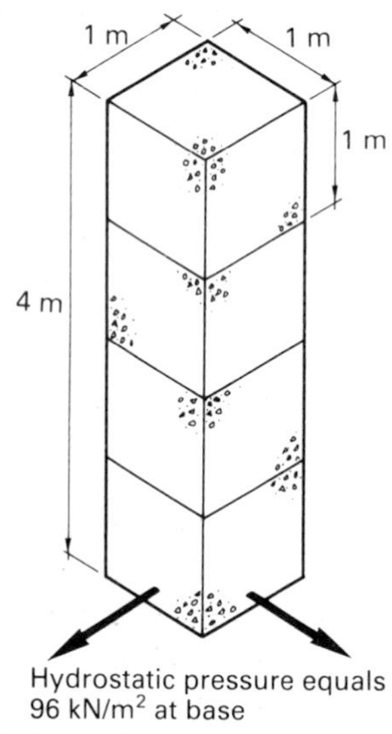

Figure 7 *Hydrostatic pressure*

This figure has to be divided by 100 to express it in kN/m^2. Therefore:

$$\text{pressure at bottom of form} = \frac{9600}{100} = 96 \text{ kN/m}^2$$

The maximum hydrostatic pressure exerted on the form face *at any point* within vertical formwork can be shown graphically as in Figure 8. The pressure at any point may be calculated using the following formula:
maximum hydrostatic pressure at any point (kN/m^2)

$$= \frac{\text{density of concrete (kg/m}^3) \times \text{distance from top of form (m)}}{100 \text{ (converts to kN)}}$$

For ease of expression this formula is simplified to:

$$P_{\text{max}} = \frac{\Delta\, H}{100} \quad \text{kN/m}^2$$

where

P_{max} = maximum hydrostatic pressure
Δ = density of concrete
H = height of pour or distance from top of form

Example
To calculate the maximum hydrostatic pressure exerted at the base of a 3 m high column form:

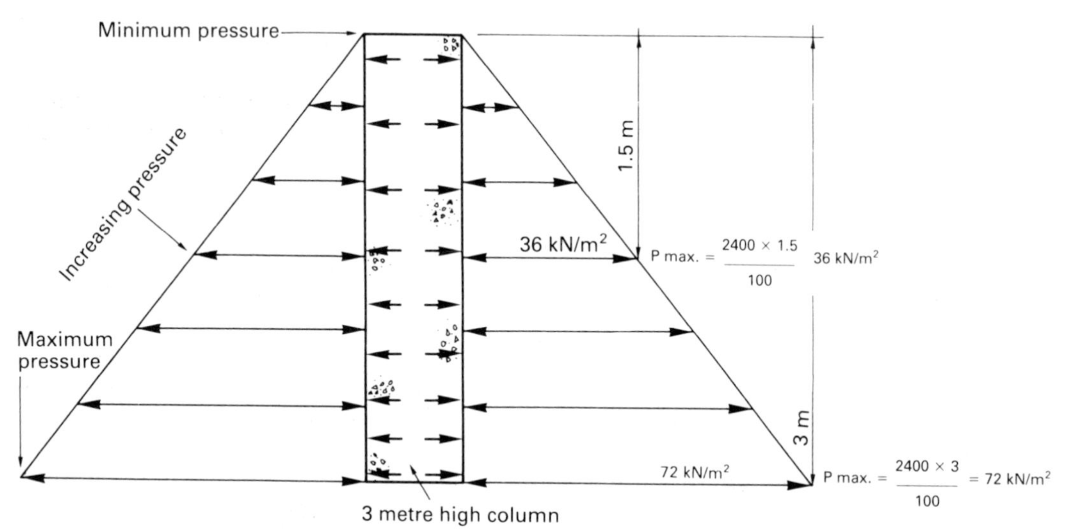

Figure 8 *Hydrostatic pressure graph*

$$P_{max} = \frac{\Delta\ H}{100} \quad kN/m^2$$

$$= \frac{2400 \times 3}{100} = 72\ kN/m^2$$

Example
To calculate the maximum hydrostatic pressure exerted halfway up a 3 m high column form:

$$P_{max} = \frac{\Delta\ H}{100} \quad kN/m^2$$

$$= \frac{2400 \times 1.5}{100} = 36\ kN/m^2$$

Factors affecting hydrostatic pressure
The method of calculating P_{max} just considered assumes that the whole of a concrete pour is completely fluid. However, in practice on site it has been found that the development of pressures within formwork is affected by the following factors:

Density of concrete Pressure increases in direct proportion to density. The greater the density of the concrete, the greater will be the pressure exerted on the form face.
Workability Pressure increases in proportion to increases in the slump of the concrete mix (see slump test, Chapter 5). A mix with a low workability (small slump) will be stiffer and thus more self-supporting than a mix with a high workability (large slump).
Rate of placing Slow rates of placing enable the lower levels of concrete to commence stiffening before the pour is complete, thus reducing pressure. On the other hand, if the rate of placing is fast the concrete at the base of the pour will still be fluid and thus at maximum pressure.
Method of discharge Discharge from a height causes a surge or impact loading which exerts greater pressure than a discharge close to the form.
Concrete temperature High temperatures quicken the stiffening and setting of concrete and thus reduce pressure. Low temperatures have the opposite effect.
Vibration As the object of vibration is to fluidize a concrete mix to force air to the surface so that a maximum compacted density is achieved, this increases the pressure up to the maximum.
Height of pour The greater the height of fluid concrete, the greater the pressure will be. In theory P_{max} increases with the height of the pour. However, in practice the lower levels of concrete will have started stiffening and setting during the time taken to fill the form, and therefore P_{max} over the full height is unlikely to be realized.
Dimensions of section Pressure in walls and columns below 500 mm thick is reduced because of an arching or support effect. The closer the form faces are together the less the pressure will be, owing to the concrete mix bridging across the gap between the formwork sides and helping to support progressively the concrete above.
Reinforcement detail Steel reinforcement tends to provide support for the concrete and therefore reduces pressure. The greater the amount of reinforcement the less the pressure will be.

Design hydrostatic pressure
In order to consider these factors the design hydrostatic pressure is taken to be the least value calculated using the following three equations, plus an addition of 10 kN/m^2 for the impact loading of the falling concrete.

Total height of pour

$$P_{max} = \frac{\Delta\ H}{100} \quad kN/m^2$$

This is the standard maximum hydrostatic pressure formula, which assumes that the entire pour is fluid.

Arching limit

$$P_{max} = 3R + \frac{d}{10} + 15 \quad kN/m^2$$

This formula takes into account the arching effect or support given to the concrete by vertical form faces less than 500 mm apart.

Stiffening of concrete

$$P_{max} = \frac{\Delta\ RK}{100} + 5 \quad kN/m^2$$

This formula takes into account the early stiffening of a mix owing to concrete temperature, workability and rate of placing.

In these equations, the symbols have the following meanings:

Δ = density of concrete
H = height of pour
R = rate of placing in metres per hour (m/h)
d = distance between vertical form faces (maximum 500 mm)
K = correction factor to allow for concrete temperature and workability of mix (see Table 8)

Table 8 *Correction factor K for hydrostatic pressure design calculations*

Workability (slump, mm)	*Concrete temperature, °C* 5	10	15	20	25	30
25	1.45	1.10	0.80	0.60	0.45	0.35
50	1.90	1.45	1.10	0.80	0.60	0.45
75	2.35	1.80	1.35	1.00	0.75	0.55
100	2.75	2.10	1.60	1.15	0.90	0.65

Example
To determine the design hydrostatic pressure for a 400 mm thick, 2.5 m high, concrete wall having a density of 2400 kg/m^3, a concrete temperature of 10°C, a slump of 50 mm and a rate of placing of 2 m/h.

Height of pour

$$P_{max} = \frac{\Delta H}{100} \quad kN/m^2$$

$$= \frac{2400 \times 2.5}{100} = 60 \ kN/m^2$$

Arching limit

$$P_{max} = 3R + \frac{d}{10} + 15 \quad kN/m^2$$

$$= (3 \times 2) + \frac{400}{10} + 15 = 61 \ kN/m^2$$

Stiffening of concrete

$$P_{max} = \frac{\Delta RK}{100} + 5 \quad kN/m^2$$

$$= \frac{2400 \times 2 \times 1.45}{100} + 5 = 74.6 \ kN/m^2$$

In this situation the height of pour equation gives the lowest value. Therefore:

design hydrostatic pressure $= 60 \ kN/m^2 + 10 \ kN/m^2 = 70 \ kN/m^2$

Example
To determine the design hydrostatic pressure for the wall in the previous example, but with a temperature of 20°C, a slump of 25 mm and a 3 m/h rate of placing.

Height of pour
As before:

$$P_{max} = 60 \ kN/m^2$$

Arching limit

$$P_{max} = 3R + \frac{d}{10} + 15 \quad kN/m^2$$

$$= (3 \times 3) + \frac{400}{10} + 15 = 64 \ kN/m^2$$

Stiffening of concrete

$$P_{max} = \frac{\Delta RK}{100} + 5 \quad kN/m^2$$

$$= \frac{2400 \times 3 \times 0.60}{100} + 5 = 48.20 \ kN/m^2$$

In this situation the stiffening of concrete equation gives the lowest value. Therefore:

design hydrostatic pressure $= 48.20 \ kN/m^2 + 10 \ kN/m^2 = 58.20 \ kN/m^2$

Pressure design graphs
Pressure design graphs are often used by the formwork designer as a speedy alternative to calculation when determining the design hydrostatic pressure. They are produced in various formats by a number of organizations, such as the

Construction Industry Research Information Association (CIRIA); formwork equipment manufacturers, for use with their proprietary systems; and even by formwork contractors for use internally within their own department.

As with the calculation method, the design hydrostatic pressure is taken to be the least of the three values obtained, plus normally an addition of 10 kN/m^2 for impact loading.

Table 9 gives the total height of pour P_{max} for a concrete mix with a density of 2400 kg/m^2. The pressure may be read off against the height of pour using the diagonal line.

Table 10 gives the P_{max} for the arching limit of concrete with a density of 2400 kg/m^2, for use where vertical form faces are less than 500 mm apart (d). The pressure is read off against the rate of placing (m/h) using the d line.

Table 11 gives the P_{max} for the stiffening of concrete. Again this is for a concrete mix with a density of 2400 kg/m^2. The pressure is read off against the rate of placing using the diagonal line, which gives the appropriate slump S (mm) and temperature (°C).

Example

To determine the design hydrostatic pressure by the use of graphs for a column 500 mm wide and 4 m in height, a temperature of 15°C, a slump of 75 mm, and a 4 m/h rate of placing.

Height of pour

From Table 9 the height of 4 m crosses the diagonal line at a pressure of 96 kN/m^2. Thus:

$$P_{max} = 96 \text{ kN/m}^2$$

Arching limit

From Table 10 a 4 m/h rate of placing crosses the d = 500 mm line at a pressure of 77 kN/m^2. Thus:

$$P_{max} = 77 \text{ kN/m}^2$$

Table 9 *Formface pressure chart (height of pour)*

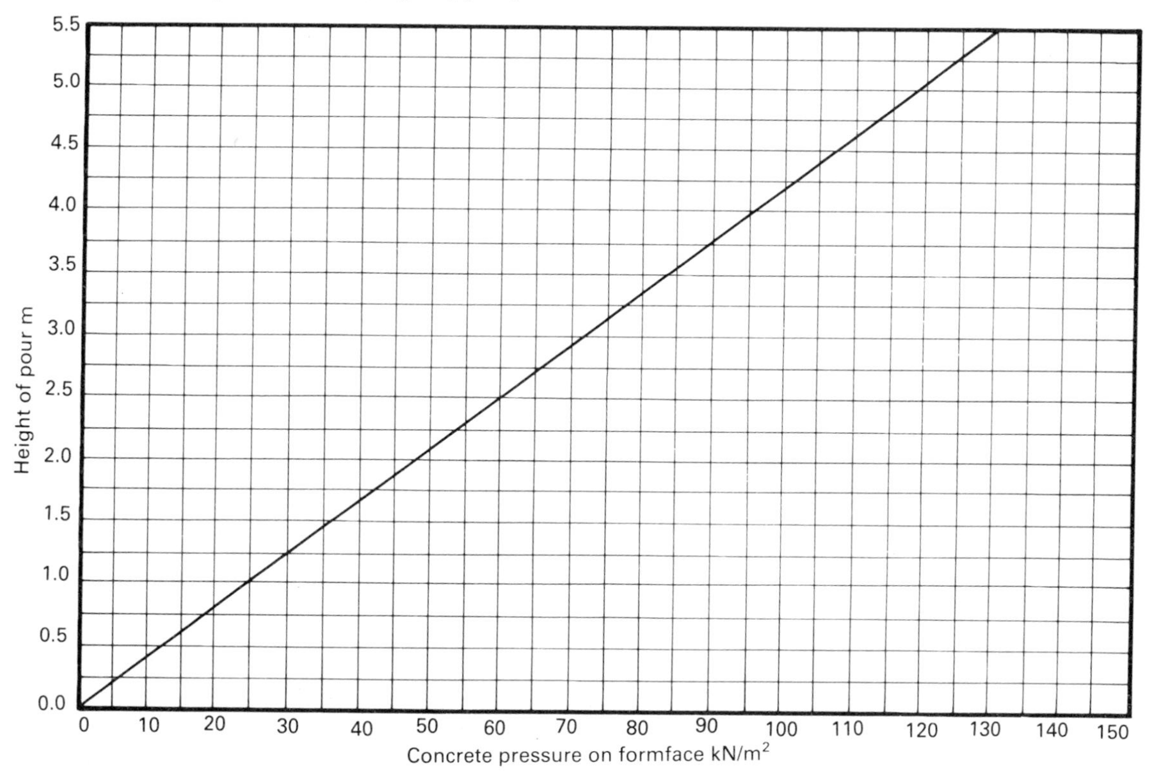

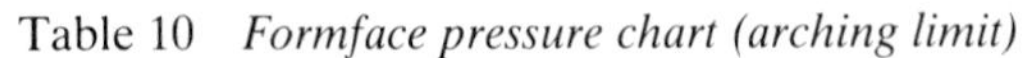

Table 10 *Formface pressure chart (arching limit)*

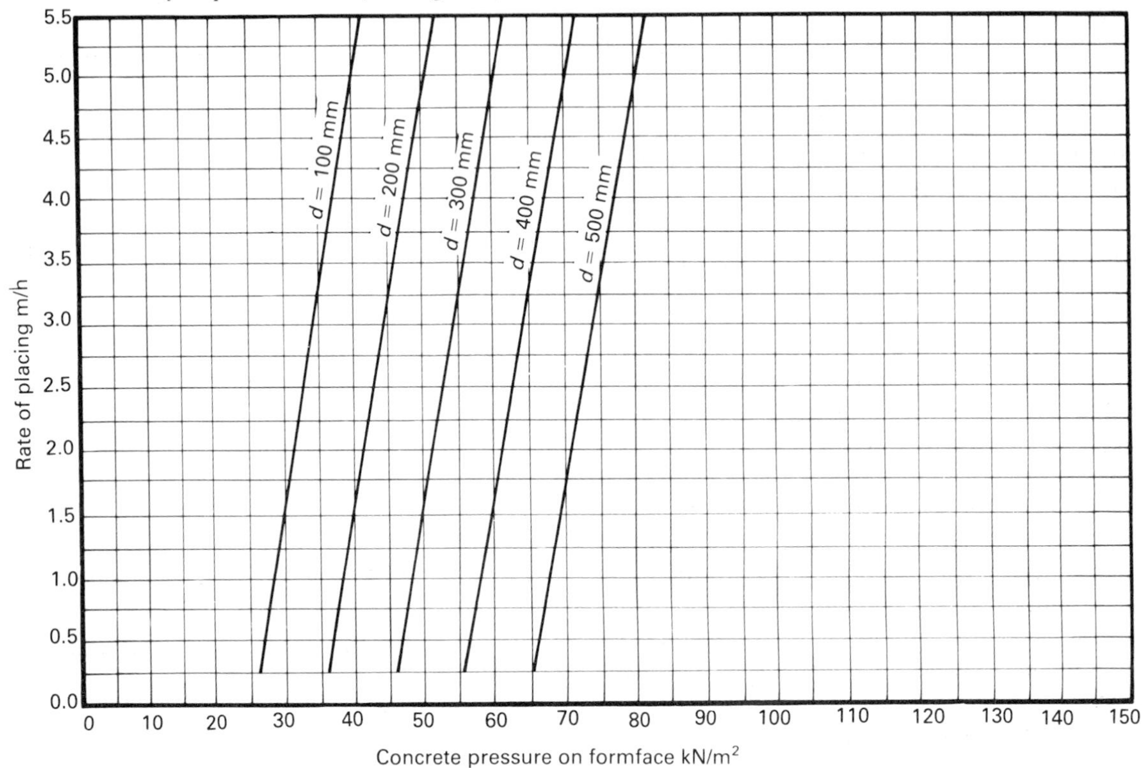

Table 11 *Formface pressure chart (stiffening)*

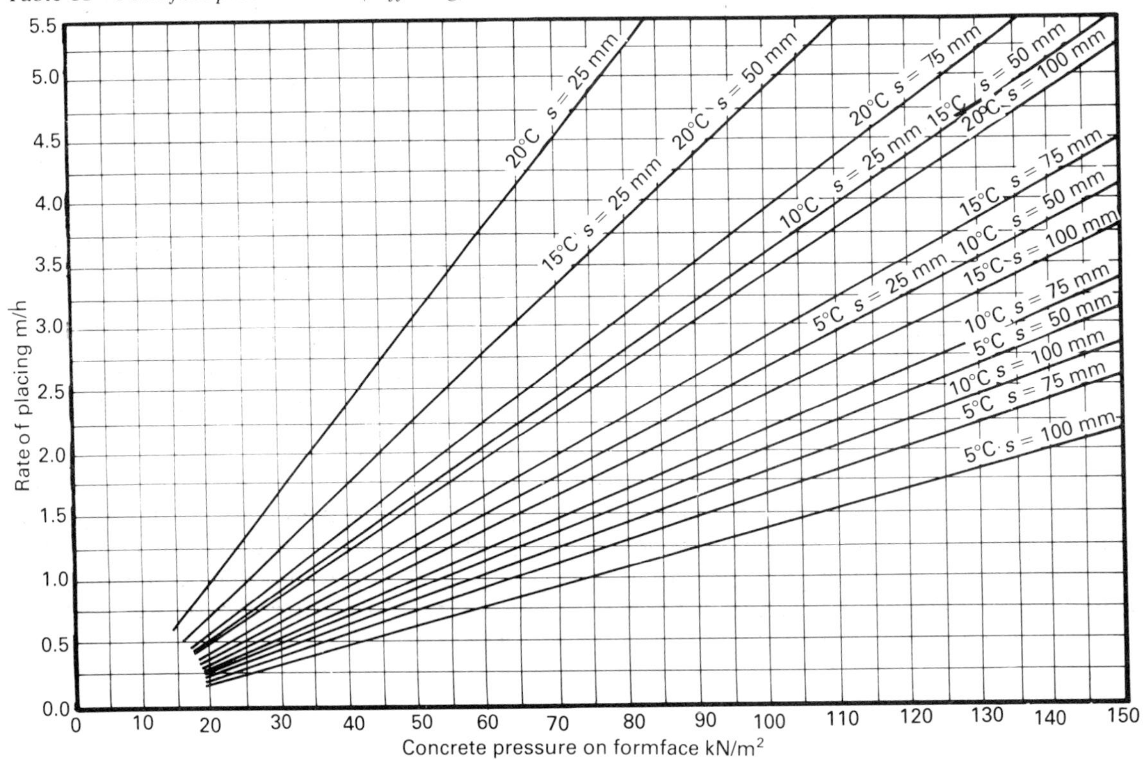

Stiffening of concrete
From Table 11 a 4 m/h of placing crosses the S = 75 mm, 15°C line at a pressure of 134 kN/m². Thus:

$$P_{max} = 134 \text{ kN/m}^2$$

In this situation the arching limit is the lowest of the three values obtained. Therefore:

design hydrostatic pressure $= 77 \text{ kN/m}^2 + 10 \text{ kN/m}^2 = 87 \text{ kN/m}^2$

Position of maximum pressure
Either the calculation method or the graph method can be used to determine the design hydrostatic pressure. Both methods give the maximum pressure on the form face. There can be assumed to be nil pressure at the top of the form, increasing to the maximum part way down the form.

The actual distance from the top of the form that the maximum pressure occurs can be determined by the use of the following formula:

$$\text{distance from top of form at which the maximum pressure occurs} = \frac{P_{max}}{24} \text{ m}$$

Note: 24 in the formula is derived by expressing the assumed concrete density in kN. For example 2400 ÷ 100 = 24.

Example
To determine the position of the maximum pressure in a 4 m high column form, where the maximum pressure is 77 kN/m² (as in previous example):

$$\text{distance from top of form at which the maximum pressure occurs} = \frac{P_{max}}{24} \text{ m} = \frac{77}{24} = 3.208 \text{ m}$$

This can be illustrated in the form of a *pressure design diagram* as shown in Figure 9. The impact loading allowance of 10 kN/m² has been included in the diagram. Thus for practical purposes the design pressure on the form face from the base of the column (point A) up to point B is 87 kN/m². It

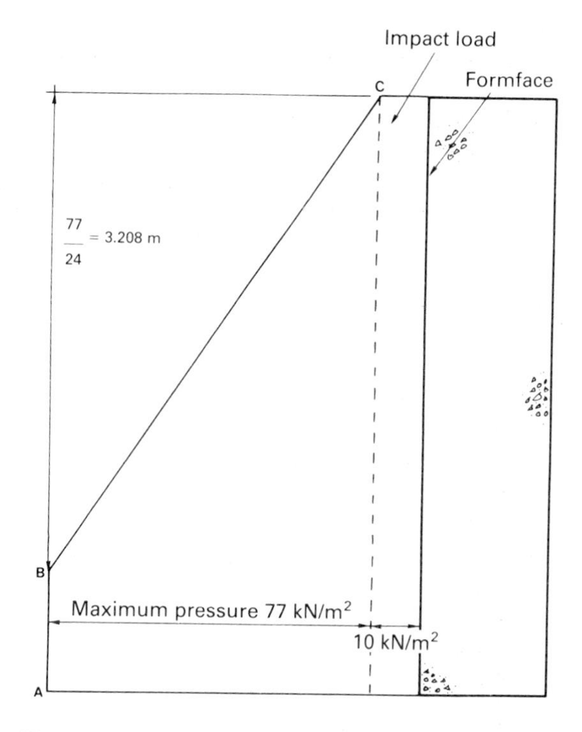

Figure 9 *Pressure design diagram*

then diminishes uniformly up to the top of the form (point C), where it is taken to be 10 kN/m².

Timber formwork design

The procedures to be followed when designing timber formwork can be considered under two distinct headings:

1 Vertical formwork design (walls, columns)
2 Horizontal formwork design (slabs).

Vertical formwork design: walls
The following procedure can be used when designing wall formwork (see Figure 10):

1 *Design pressure*
Determine the design hydrostatic pressure in kN/m² (least value of P_{max} from height of pour, arching limit, or stiffening of concrete, plus an impact loading allowance of 10 kN/m²), either by calculation or by use of graphs (Tables 9, 10, 11).

2 *Plywood*
Decide on the type of plywood. Thickness 18/19 mm plywood is used almost universally for formwork, as it is the most cost effective thickness in terms of both actual plywood cost and spacing of framing members.

Use pressure from 1 and Table 12 to fix the span of the plywood. Make this a modular fraction of the sheet size, e.g. approximately 200 mm, 300 mm or 400 mm. This will be the spacing used for the framing members.

3 *Framing*
Use pressure from 1 and plywood span from 2 to calculate load per metre run on framing members:

kN/m run on framing member
$= \text{design } P_{max} \times \text{plywood span}$

From Table 13 determine a suitable section and span for the framing. The framing span will be the spacing for the soldiers or walings. This is normally taken to be between one and two times the framing spacing, and should be a modular fraction of either the form length for soldiers, or the form height for walings.

4 *Soldier or waling*
Use pressure from 1 and framing spacing from 3 to calculate load per metre run on the soldiers or walings:

kN/m run on soldier or waling
$= \text{design } P_{max} \times \text{framing spacing}$

Divide by two when soldier or walings are used in pairs, and from Table 13 determine a suitable section and span for the soldiers or walings. The soldier or waling span will be the spacing in one direction for the form ties; the other will be the framing span.

5 *Form ties*
Use pressure from 1, framing span from 3 and soldier or waling span from 4 to calculate the form tie load:

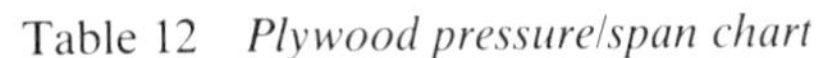
Table 12 *Plywood pressure/span chart*

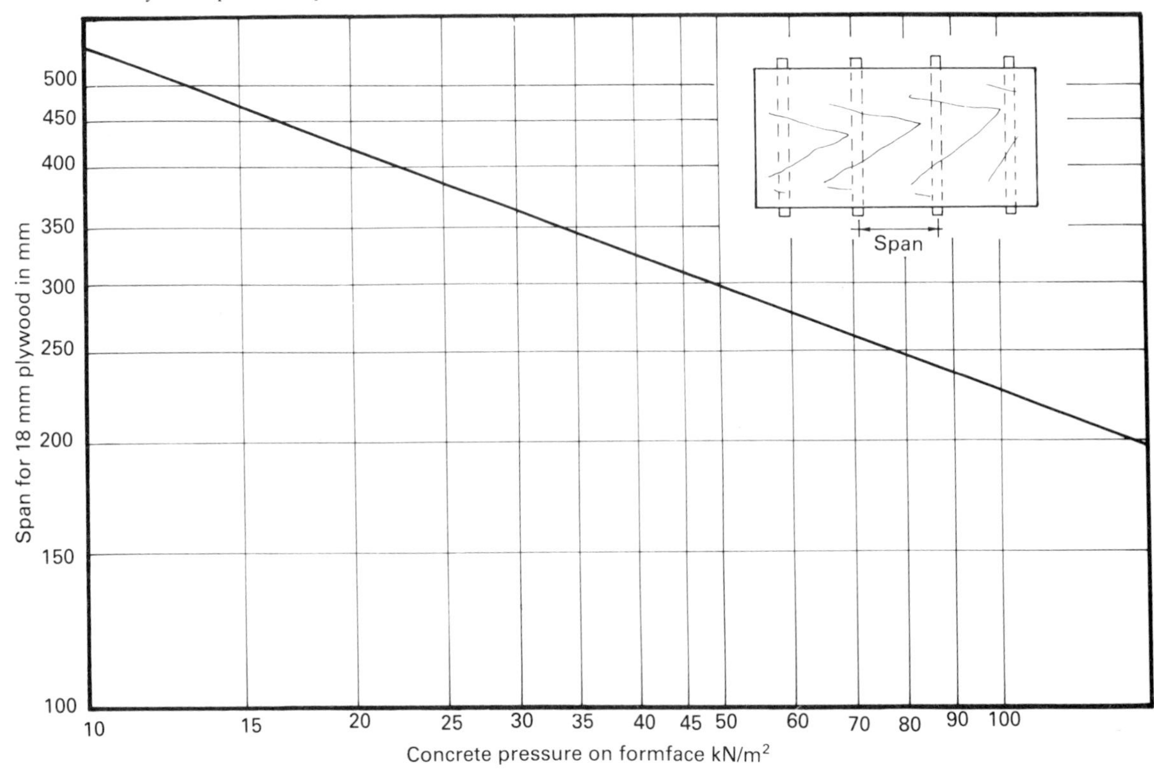

form tie load, kN
= design P_{max} × framing span × soldier or waling span

and from the manufacturer's literature select size and type of form tie.

6 *Redesign*
Repeat previous design procedure if required using different spacings, sectional sizes, or form ties to determine alternative solutions. Select the most economical.

Example
To determine formwork details for a 150 mm thick by 2.5 mm high wall form with horizontal framing and twin vertical soldiers, using the following:

Concrete density	2400 kg/m^3
Slump	75 mm
Temperature	15°C
To be filled in	1 hour

1 *Design pressure*
(a) Height of pour: from Table 9, P_{max} = 60 kN/m^2.
(b) Arching:

$$\text{rate of placing, m/h} = \frac{\text{height}}{\text{filling time}}$$

Table 13 *(a) strength class 4 (decking members); (b) strength class 4 (column members); (c) strength class 4 (wall members)*

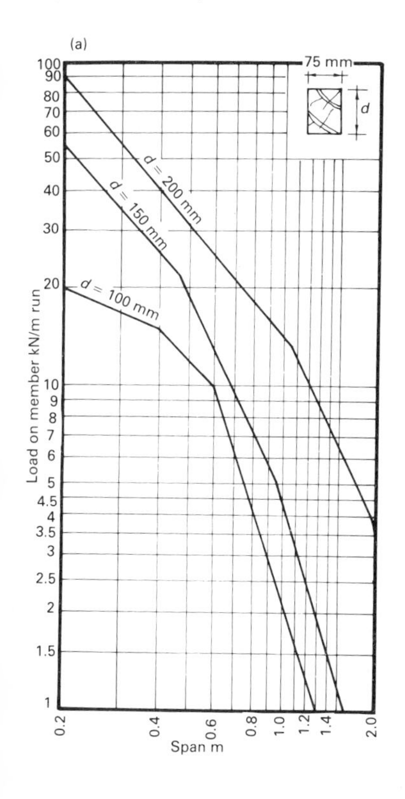

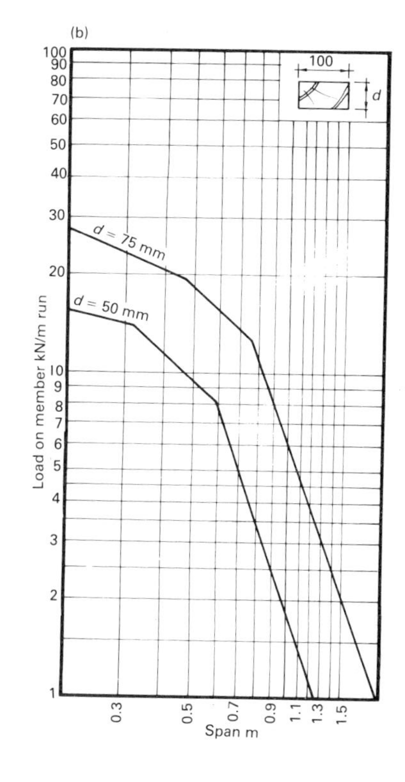

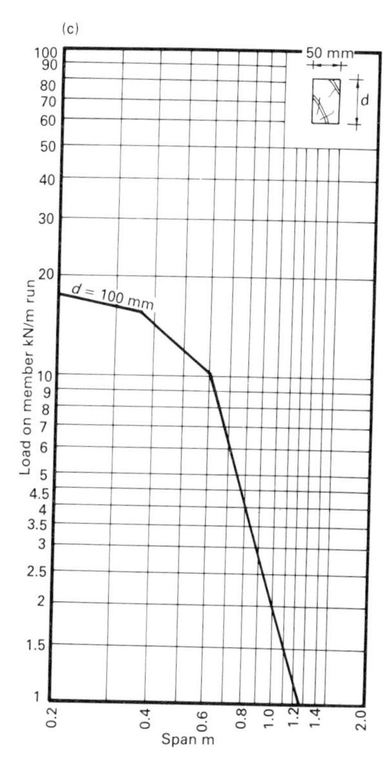

$$= \frac{2.5}{1} = 2.5 \text{ m/h}$$

From Table 10, 2.5 m/h crosses $d = 150$ mm at $P_{max} = 38$ kN/m^2.

(c) Stiffening: from Table 11, 2.5 m/h crosses $S = 75$ mm, 15°C line at $P_{max} = 85$ kN/m^2.

Thus:
design $P_{max} = 38$ kN/m^2 + 10 kN/m^2 = 48 kN/m^2

2 *Plywood*
From Table 12, load of 48 kN/m^2 gives a span of 300 mm.

3 *Framing*
kN/m run on framing = 48 × 0.3 = 14.4 kN/m run

From Table 13, use 50 mm × 100 mm at 400 mm span.

4 *Soldiers*
kN/m run on soldiers = 48 × 0.4 = 19.2 kN/m run

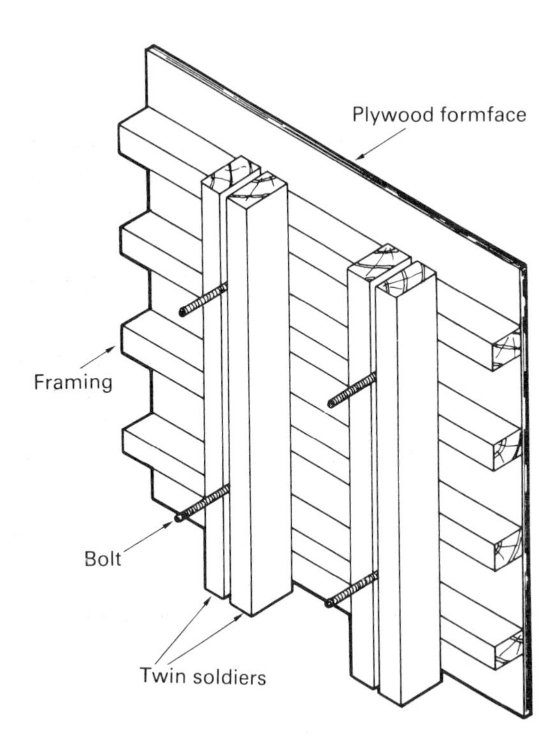

Figure 10 *Wall formwork design*

Divide by two for twin soldiers:

$$\text{kN/m run on one soldier} = \frac{19.2}{2} = 9.6 \text{ kN/m run}$$

From Table 13, use 50 mm × 100 mm at 600 mm span.

5 *Form ties*
Form tie load, kN = 48 × 0.4 × 0.6 = 11.52 kN

6 *Redesign*
To provide an alternative solution if required.

Vertical formwork design: columns

The following procedure can be used when designing column forms (see Figure 11):

1 *Design pressure*
Determine the design hydrostatic pressure in kN/m^2 (least value of P_{max} from height of pour, arching limit, or stiffening of concrete, plus an impact loading allowance of 10 kN/m^2), either by calculation or by use of graphs (Tables 9, 10, 11).

2 *Plywood*
Use pressure from 1 and Table 12 to determine the maximum span of plywood. This will be the spacing of the vertical backing timbers.

3 *Backing timber*
A backing timber fixed to each edge of the plywood column side is normally suitable; large section or tall columns may require additional ones to come below the maximum plywood span. Use pressure from 1, width of plywood side and number of backing timbers to calculate load per metre run on each backing timber:

kN/m run on each backing timber

$$= \frac{\text{design } P_{max} \times \text{width of ply side}}{\text{no. of backing timbers}}$$

and from Table 13 determine a suitable section and maximum span for the backing timbers. Check centre to centre spacing of backing timbers does not exceed plywood span.

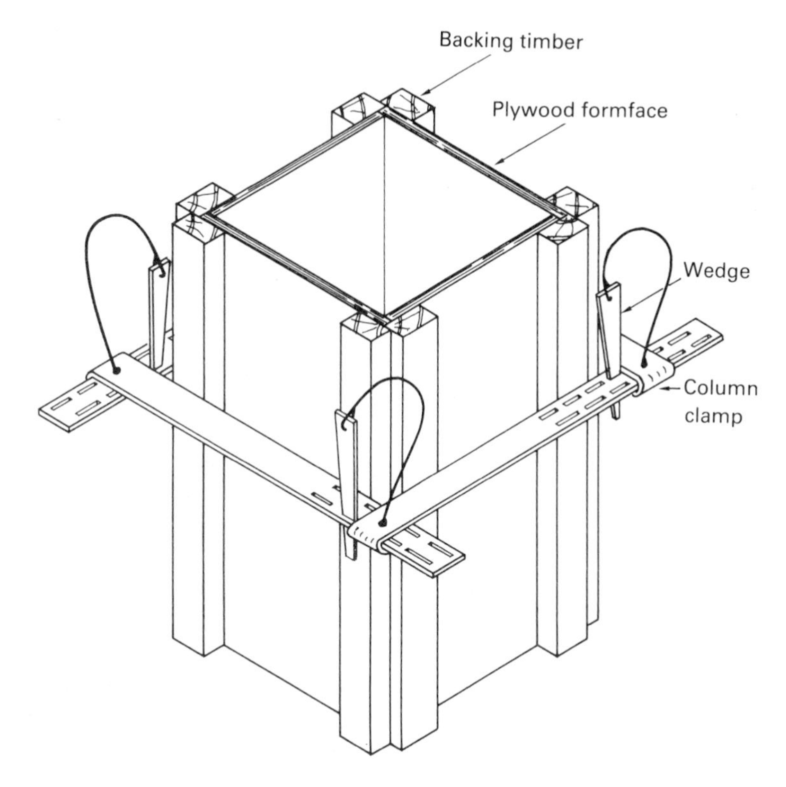

Figure 11 *Column formwork design*

4 *Column clamp positions*
Draw pressure diagram to scale and assume clamp positions (see Figure 12). Using maximum span from 3 as clamp spacing up to point of maximum pressure, increasing each subsequent position by 50 mm above this point.
(a) Measure pressure at each clamp position.
(b) Multiply pressure at each clamp position by width of ply side and divide by number of backing timbers to obtain load on each backing timber at each clamp position:

kN on each backing timber at clamp position

$$= \frac{\text{pressure at clamp} \times \text{width of ply side}}{\text{no. of backing timbers}}$$

(c) Starting from the lower clamp, check, using Table 13 and answers obtained in (b), that the distance between it and the next clamp does not exceed the span for the section of backing timber. If it does, reduce the spacing of clamps appropriately and recheck.

Note This method errs on the safe side, as it assumes that the pressure between two clamps is equal to the lower one throughout.

5 *Redesign*
Repeat previous design procedure if required using different spacings for clamps, sectional sizes and number of backing timbers, to determine alternative solutions. Select the most economical.

Example
To determine formwork details for a 300 mm square column 3.0 m high, using the following:

Concrete density	2400 kg/m^3
Slump	50 mm
Temperature	15°C
To be filled in	0.75 hour

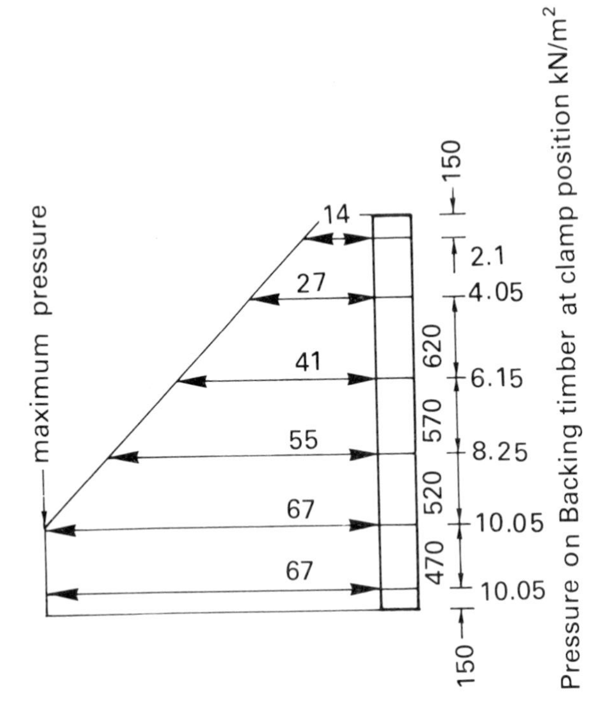

Figure 12 *Column formwork (clamp positions) design example*

1 *Design pressure*

(a) Height of pour: from Table 9, P_{max} = 72 kN/m^2.

(b) Arching:

$$\text{rate of placing, m/h} = \frac{\text{height}}{\text{filling time}}$$

$$= \frac{3}{0.75} = 4 \text{ m/h}$$

From Table 10, 4 m/h crosses d = 300 mm at P_{max} = 57 kN/m^2.

(c) Stiffening: from Table 11, 4 m/h crosses S = 50 mm, 15°C at P_{max} = 110 kN/m^2.

Thus:

design P_{max} = 57 kN/m^2 + 10 kN/m^2 = 67 kN/m^2

2 *Plywood*

From Table 12, load of 67 kN/m^2 gives a span of 270 mm.

3 *Backing timber*

kN/m run on each backing timber

$$= \frac{67 \times 0.3}{2} = 10.05 \text{ kN/m run}$$

From Table 13, 50 mm × 100 mm has a maximum span of 470 mm. Backing timbers of 50 mm × 100 mm give a plywood span of approximately 220 mm, which is less than maximum plywood span (see Figure 13).

4 Draw pressure design graph and assume clamp positions.

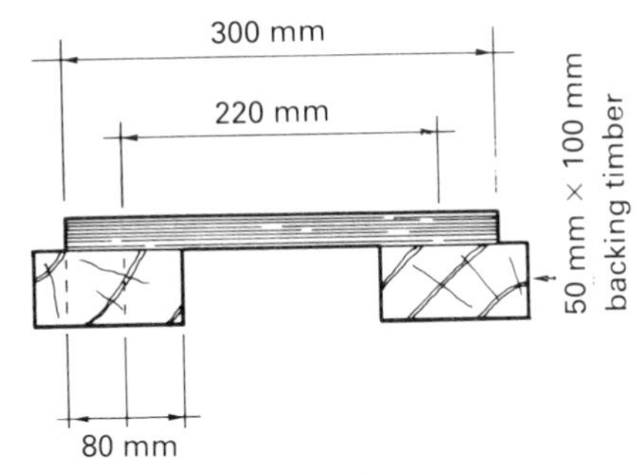

Figure 13 *Column plywood span*

5 *Redesign*

To provide an alternative solution if required.

Horizontal formwork design: slabs

The following procedure can be used when designing slab formwork (see Figure 14):

1 *Design loading*

Determine the design loading in kN/m^2:

design load = (24 × slab thickness) + imposed load allowance

Take imposed load to be:

3.5 kN/m^2 for decking

2 kN/m^2 for other structural members (joists, ledgers, props).

2 *Plywood*

Use design load from 1 and Table 12 to fix the span of the plywood. As with walls, make this a modular fraction of the sheet size. This will be the spacings of the joists.

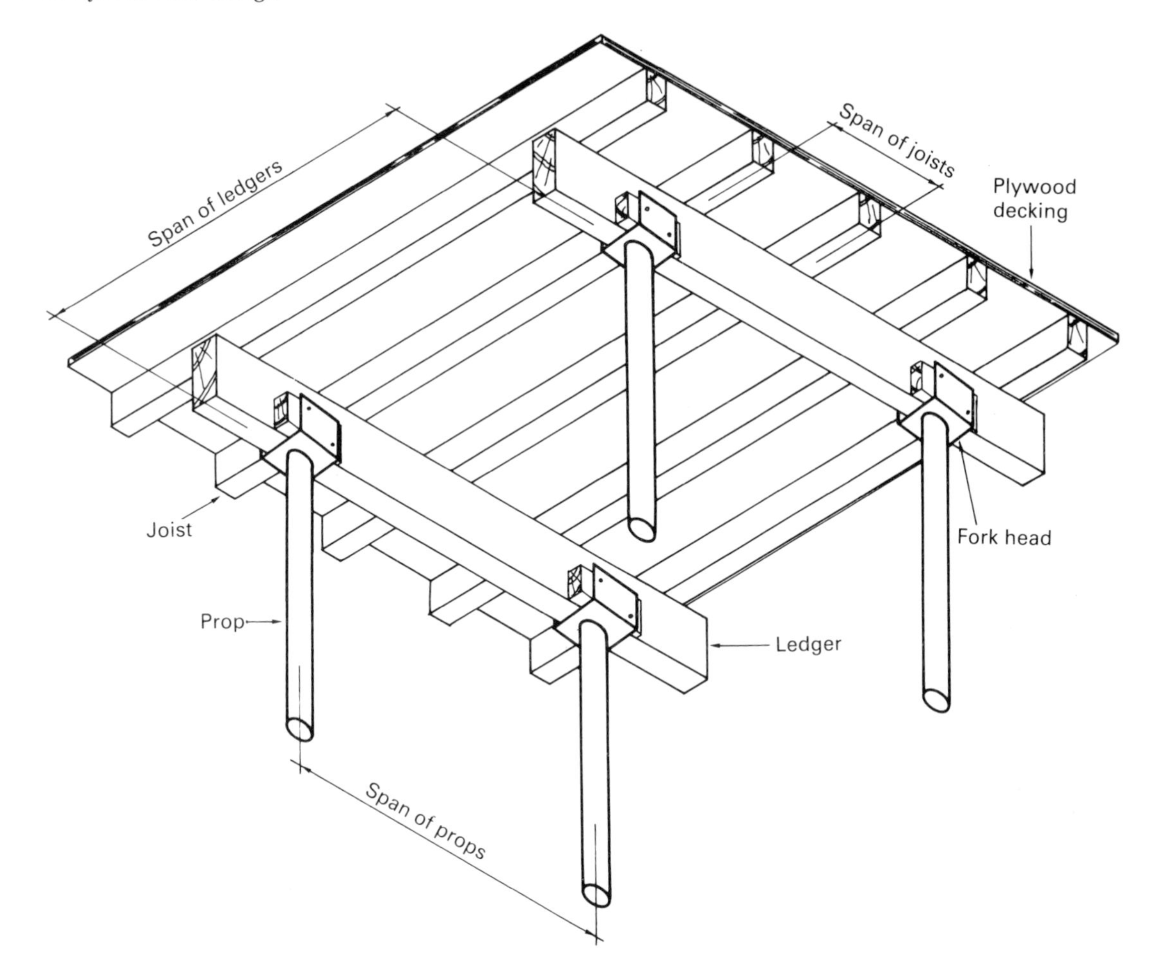

Figure 14 *Slab formwork design*

3 *Joists*
Use design load from 1 and plywood span from 2 to calculate load per metre run on joists.

kN/m run on joists = design load × plywood span

and from Table 13 determine a suitable section and span for the joists. This span will be the spacings for the ledgers.

4 *Ledgers*
Use design load and ledger spacing from 3 to calculate the load per metre run on the ledgers:

kN/m run on ledger = design load × ledger spacing

and from Table 13 determine a suitable section and span for the ledgers. The ledger span will be the spacing in one direction for the adjustable steel props. The other will be the joist span.

5 *Adjustable steep props*
Use design load from 1, joist span from 3 and ledger span from 4 to give the load on one adjustable steep prop:

prop load, kN
= design load × joist span × ledger span

Determine extended prop height:

extended prop height
= storey height − ply − joist depth − ledger depth

From Table 14, which gives typical adjustable steel prop safe loadings with or without lacings, select suitable props. Where prop loading is too great, joist or ledger span must be reduced accordingly.

6 *Redesign*
Repeat previous design procedure if required using different spans or sectional sizes, to determine alternative solutions. Select the most economical.

Example
Determining timber formwork details for a 300 mm thick slab which has a storey height (support surface to underside of soffit) of 3.5 m. Assume the use of unlaced props.

1 *Design loading*

design load, decking = $(24 \times 0.3) + 3.5 = 10.7$ kN/m^2
design load, supports = $(24 \times 0.3) + 2 = 9.2$ kN/m^2

2 *Plywood*
From Table 12, a load of 10.7 kN/m^2 gives a maximum span of 460 mm. Thus use 400 mm for joist spacing.

3 *Joists*

kN/m run on joists = $9.2 \times 0.4 = 3.68$ kN/m run

From Table 13, use 75 mm × 100 mm to span 1 m.

4 *Ledgers*

kN/m run on ledger = $9.2 \times 1 = 9.2$ kN/m run

From Table 13, use 75 mm × 150 mm to span 1.4 m.

5 *Adjustable steel props*

prop load, kN = $9.2 \times 1 \times 1.4 = 12.88$ kN
extended prop height
= $3.5 - 0.018 - 0.1 - 0.15 = 3.232$ m

From Table 14(a), prop sizes 2, 3 and 4 are suitable for length and capable of carrying the load.

6 *Redesign*
To provide an alternative solution if desired.

Table 14 *Typical safe working loads (kN) for concentrically loaded props*
(a) Without lacings

Maximum prop height m	*Prop size 1*	*2*	*3*	*4*
1.750	32	*	*	*
2.000	32	32	*	*
2.250	32	32	*	*
2.500	26	26	*	*
2.750	23	23	23	*
3.000	19	19	19	*
3.250	*	17	17	24
3.500	*	*	15	19
3.750	*	*	13	15
4.000	*	*	*	12
4.250	*	*	*	11
4.500	*	*	*	10
4.750	*	*	*	9

*Prop not suitable
Note Props for formwork are commonly available in the four standard sizes shown. These have the following height ranges:
1 1.75–3.12 m
2 1.98–3.35 m
3 2.59–3.96 m
4 3.2–4.87 m

(b) With lacings

Maximum prop height m	*Prop size 1*	*2*	*3*	*4*
2.500	32	32	*	*
2.750	32	32	32	*
3.000	32	32	32	*
3.250	*	32	32	32
3.500	*	*	28	32
3.750	*	*	24	30
4.000	*	*	*	26
4.250	*	*	*	22
4.500	*	*	*	19
4.750	*	*	*	16

*Prop not suitable

Alternative to timber design graphs
Calculations can be carried out as an alternative to the use of design graphs when determining the sectional size of timber formwork members.

The sectional size of formwork structural members has to be designed so that the permissible bending stress for the strength class of the timber being used is not exceeded. The following information is given only as a basic introduction to structural timber design calculations. Detailed structural design of formwork is the province of specialist formwork structural designers and engineers.

The equation for determining the sectional size of timber members is:

$$\text{maximum bending moment } BM_{max} = \text{moment of resistance } MR$$

Maximum bending moment This is the total sum of the forces acting on a timber member that tend to bend or break it. The actual BM_{max} of a member will vary according to its loading condition and means of support. Table 15 contains typical situations.

Moment of resistance This is the resistance to bending offered by a timber member's sectional size. For rectangular section timber the moment of resistance formula is:

$$MR = \frac{fbd^2}{6}$$

where

f = permissible bending stress (N/mm^2)
b = breadth of section (mm)
d = depth of section (mm)

Table 16 shows typical values for the permissible bending of stress graded softwood. These values are dependent on the strength class (SC) (see Table 1) of the timber being used. The values indicated have been adjusted to take into account the use of green or wet timber (timber with a moisture content (MC) above 18 per cent) and the temporary nature of the loading.

It is normal to determine a suitable sectional size for a given span and spacing. In cases where the calculated sectional size is found to be uneconomical, the span and or spacings should be reduced and the sectional size recalculated.

Example
To determine a suitable sectional size for softwood ledgers in slab formwork, spaced at 1.5 m centres and spanning 1.2 m, using strength class 3 timber and a design load of 6.8 kN/m^2.

Table 15 *Loading conditions*

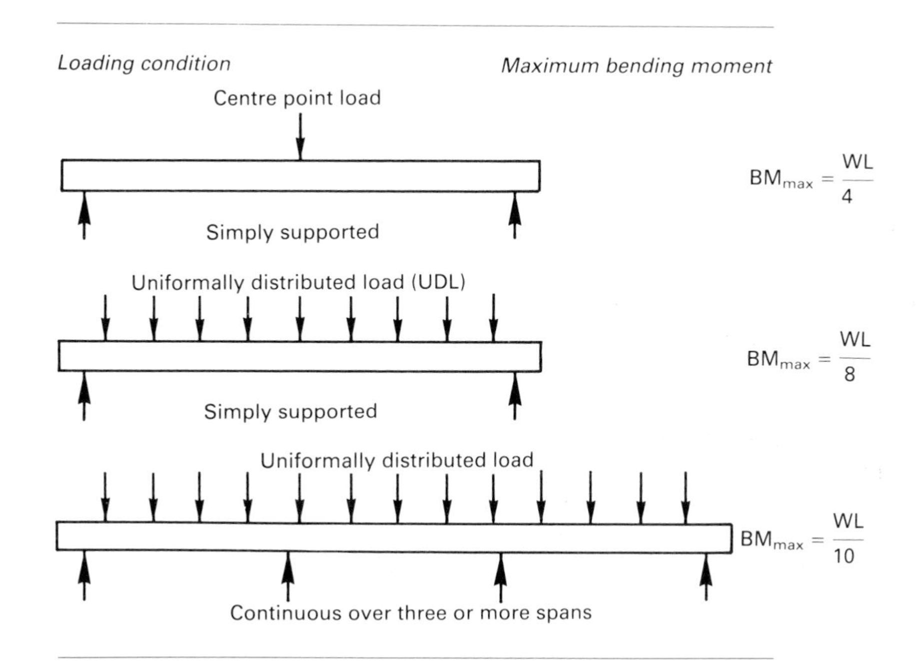

Table 16 *Typical permissible grade stresses for wet formwork members*

Strength class	*Bending stress* N/mm^2
SC1	3.1
SC2	4.6
SC3	5.9
SC4	8.4
SC5	11.2

Determine the load on a ledger between supports:

load = design load × ledger span × ledger spacing
= 6.8 × 1.2 × 1.5 = 12.24 kN

Now determine the sectional size. Use $WL/10$ for BM_{max} from Table 15 (although the ledger is subjected to a series of point loads from the joist, it is normally considered to be uniformly distributed). Therefore:

$$BM_{max} = MR$$

$$\frac{WL}{10} = \frac{fbd^2}{6}$$

Transpose the formula to get bd^2 on its own:

$$bd^2 = \frac{6WL}{10f}$$

$$= \frac{6 \times 12\,240 \times 1200}{10 \times 5.9}$$

$$= 1\,493\,694.9$$

Say b = 75 mm. Then:

$$d = \sqrt{\left(\frac{1\,493\,694.9}{75}\right)} = 141 \text{ mm}$$

Therefore the nearest commercial size (NCS) (from Figure 1) is 75 mm × 150 mm.

Notes

1 All measurements must be stated in the same units, e.g. change m to mm and kN to N by multiplying by 1000.

2 Anything can be moved from one side of an equals sign to the other, providing that it either changes its symbol or moves diagonally across from the top to the bottom or the bottom to the top. For example, plus changes to minus; multiplication changes to division; powers change to roots; that on the top line moves diagonally to the bottom line; and conversely that on the bottom line moves diagonally to the top line.

Example

To determine a suitable depth for 50 mm breadth wall form framing members, spaced at 400 mm centres and spanning 600 mm, using strength class 4 timber and a design P_{max} of 46 kN/m^2.

Determine the load on a single span of a framing member:

load = design P_{max} × framing span × framing spacing
= 46 × 0.6 × 0.4 = 11.04 kN

Now determine the sectional size. Use $WL/10$

$$BM_{max} = MR$$

$$\frac{WL}{10} = \frac{fbd^2}{6}$$

Transpose the formula to get d on its own:

$$d = \sqrt{\left(\frac{6WL}{10fb}\right)}$$

$$= \sqrt{\left(\frac{6 \times 11\,040 \times 600}{10 \times 8.4 \times 50}\right)} = 97 \text{ mm}$$

Therefore the NCS (from Figure 1) is 50 mm × 100 mm.

Example

Check to determine whether or not the following design for beam form ledgers is satisfactory using SC4 timber. Two 75 mm × 100 mm ledgers are spaced at 600 mm centres and simply supported at either end by adjustable steel props. The distance between the props is 1.2 m and the total design load to be carried by each ledger is 7.5 kN.

Use $WL/8$ for BM_{max} (from Table 15):

$$BM_{max} = MR$$

$$\frac{WL}{8} = \frac{fbd^2}{6}$$

As f is to be checked, transpose the formula to get f on its own:

$$f = \frac{6W.L}{8bd^2}$$

$$= \frac{6 \times 7500 \times 1200}{8 \times 75 \times 100 \times 100} = 9 \text{ N/mm}^2$$

From Table 16 the permissible bending stress for SC4 timber is 8.4 N/mm^2. Therefore either SC5 would have to be used, or the sectional size of SC3 would have to be increased to say 75 mm × 150 mm.

Recheck using 75 mm × 150 mm SC3:

$$f = \frac{6 \times 7500 \times 1200}{8 \times 75 \times 150 \times 150} = 4 \text{ N/mm}^2$$

Thus from Table 16 this sectional size would be suitable for bending stress using either SC2, SC3 or SC4 timber.

Self-assessment questions

Question *Your answer*

1 State the *two* loads that must be considered when determining the design load for a floor slab.

____

____

____

____

2 Name the *three* methods used to determine the design hydrostatic pressure for vertical formwork.

____

____

____

____

3 Name *three* factors that influence the development of hydrostatic pressure within vertical formwork, and for *each* state its effect.

____

____

____

____

4 The addition made to P_{max} to allow for the impact loading of falling concrete into vertical forms is:
(a) 2 kN/m^2
(b) 3.5 kN/m^2
(c) 7.5 kN/m^2
(d) 10 kN/m^2

a [ ]	b [ ]	c [ ]	d [ ]

5 The distance from the top of a 3 m high column form at which the maximum pressure of 48 kN/m^2 occurs is:
(a) 2 m
(b) 3 m
(c) 8 m
(d) 16 m

a [ ]	b [ ]	c [ ]	d [ ]

6 Briefly outline the procedure to follow when determining formwork design details for timber-framed ply-faced wall formwork.

7 Determine a suitable sectional size for decking joists spaced at 400 mm centres and spanning 600 mm, assuming a design load of 36 kN/m^2. Use

$WL/10 = fbd^2/6$, where $f = 4.6$ N/mm^2.

Chapter 3

Formwork construction

The formwork details considered in this chapter are those which may be regularly encountered in the general range of construction work. These mainly consist of timber and plywood construction, secured or supported with ancillary proprietary items. The more specialist proprietary formwork details for civil engineering work and large scale constructions are not considered. These vary widely from one manufacturer to another, who will in any event supply technical details and manuals which give erection, striking and maintenance procedures for their own particular system.

Foundations

In firm subsoil it is often possible to excavate foundations to the required depth and subsequently cast the concrete against the excavated faces. In situations where this is not possible, formwork will be required. The degree of accuracy and standard of finish for foundation formwork is not normally as high as that required for the superstructure (walls, columns, floor slabs etc.). As such it is often possible to use reclaimed materials from other sites to construct economic foundation forms.

Figure 15 illustrates typical formwork details suitable for shallow concrete strip, raft, ring beam or pad foundations. Its construction takes the form of 50 mm × 100 mm framed panels on to which is fixed 18 mm plywood, which are wedged to line, levelled and subsequently strutted in position. The strutting in level ground may be taken either from stakes at about 600 mm spacing or off a well bedded pegged sole plate. Care should be taken to avoid strutting too far up the stakes as movement will occur. For best results strutting should be as low as possible. Alternatively in deeper excavations the strutting may be taken off a sole plate laid against the bank.

In situations where blinding concrete has not been laid a polythene sheet is sometimes laid over the foundation base and extending up the form sides. As well as protecting the steel reinforcement from contamination by mud, it also helps to prevent grout loss under the edge forms. The distance pieces which maintain the correct spacing between the edge forms should be removed during pouring as the concrete reaches their respective levels.

Figure 16 again illustrates typical foundation formwork details. These are suitable for deeper column pad bases or pile caps. Framed form panels are used for both the sides and the ends. They are wedged to level and then firmly strutted in position, again from either stakes, a sole plate or off the bank. If care is taken during the laying of the blinding concrete to form it to the required plan shape it can be used as a kicker, thus positioning the form and preventing grout leakage. Alternatively as before the form sides and base may be lined with polythene sheeting.

The strutting shown in the previous details has been either cut square at both ends and tightened with folding wedges (do not forget to drive a nail through the wedges to prevent them slackening off), or bird's mouthed around a member at one end and cut square at the other. When using the latter method the strut is tightened and conse-

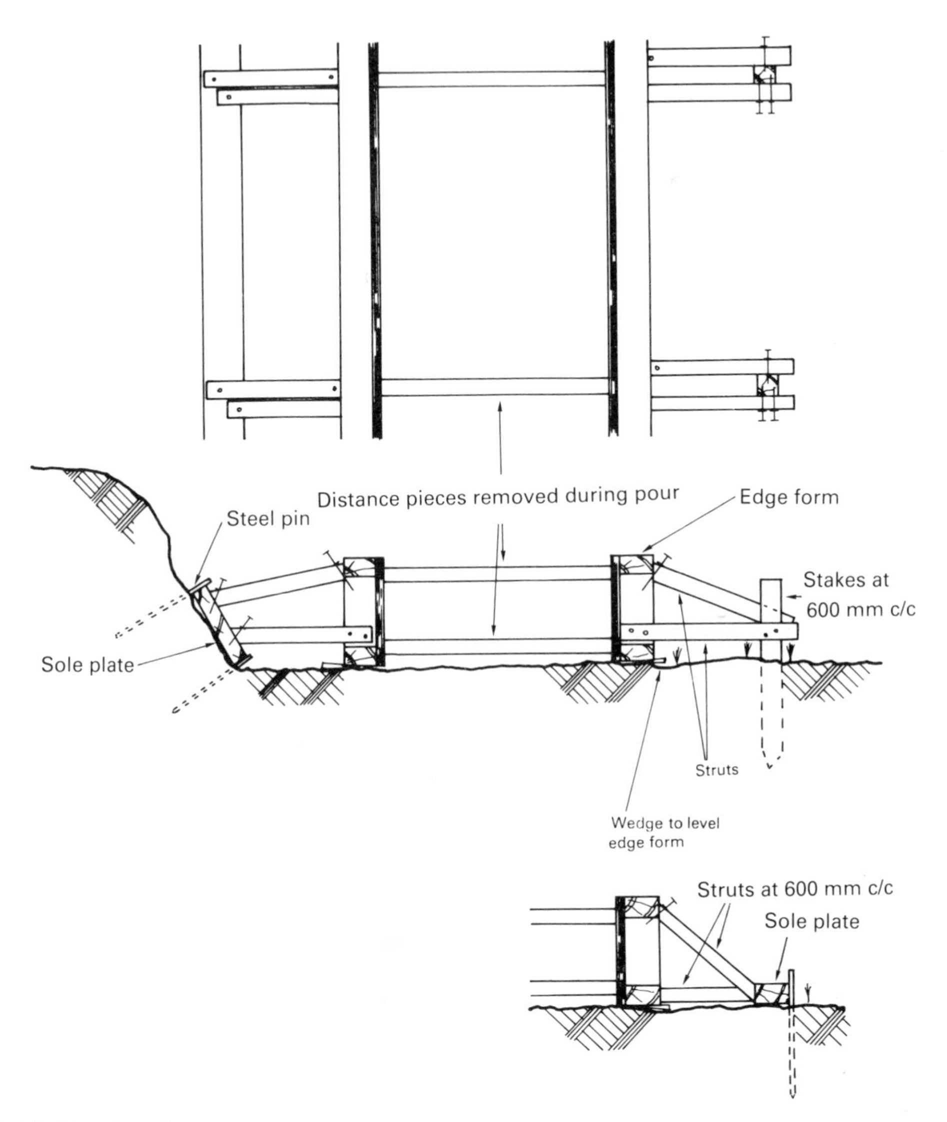

Figure 15 *Shallow foundation*

quently the form is moved into line or plumb by hammering the square end forward, causing the timbers to bite into each other and create a satisfactory bearing. This should also be secured with a nail (double headed or head left protruding for ease of withdrawal) to prevent movement.

Also shown in Figure 16 is a method which is used to support and locate the column steel reinforcement starter bars. This consists of a simple framework suspended across the form sides and notched around the vertical reinforcement.

Often pad bases have to be provided with protruding hold-down bolts or bolt pockets for the later fixing of structural columns or portal frames. A simple framework similar to that used for locating starter bars can also be used for supporting and locating these hold-down bolts. The bolts are normally required to have a certain amount of tolerance movement in order to

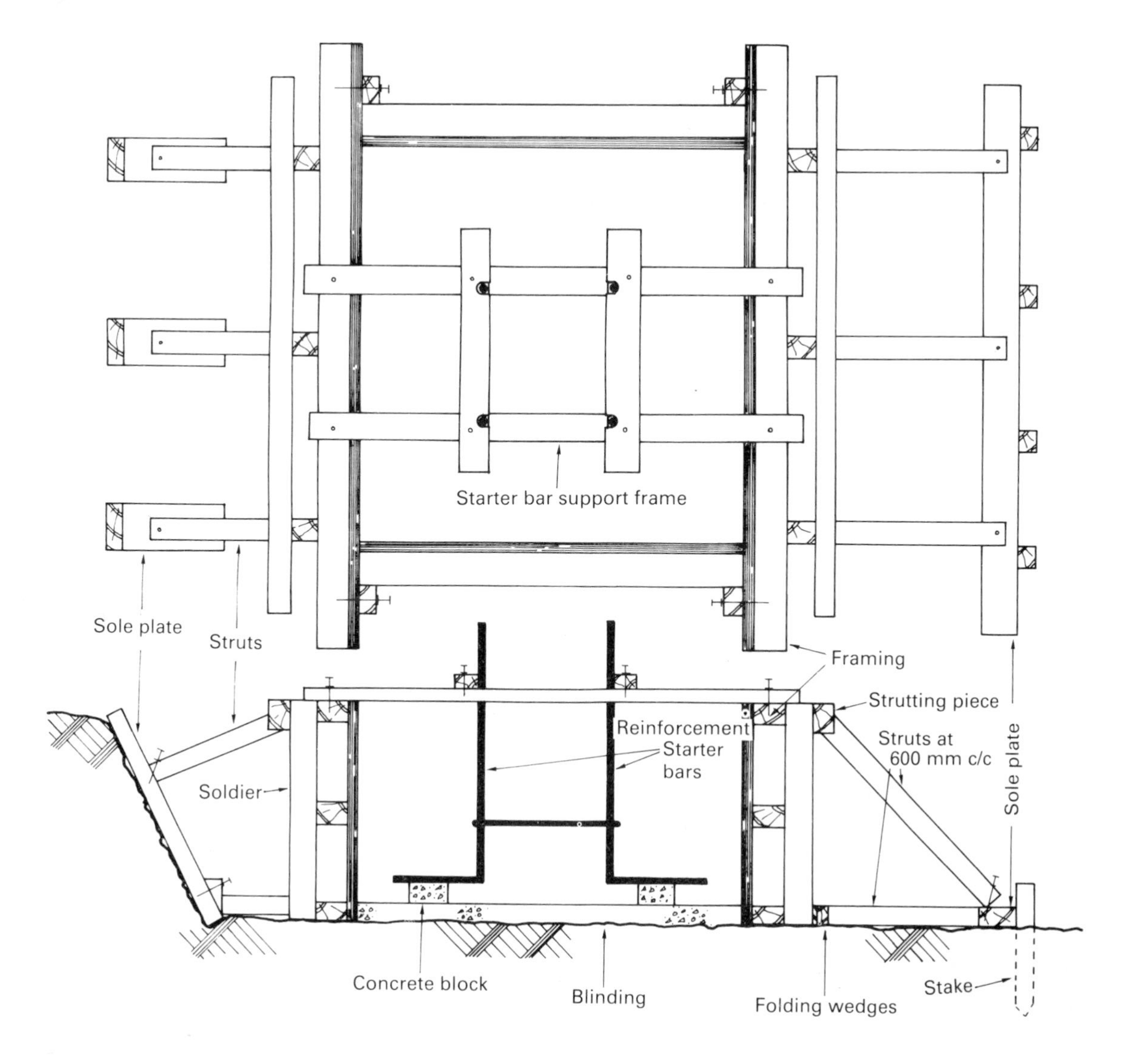

Figure 16 *Column pad base formwork*

permit the later accurate positioning of the structural framework. This can be achieved as shown in Figure 17 by the use of tapered polystyrene sleeves around the bolts which can be burnt or melted out after the concrete has been cast.

In situations where several pad bases or pile caps with the same plan size but of different depth are required, the formwork can be constructed to the greatest depth. When these are subsequently used for the shallower pours, the concrete casting level can be indicated, preferably by battens fixed to the sides. A line of nails along the form sides is often used, but this method is frowned upon by the concreters who may damage their hands on them when trowelling off.

Proprietary road edge forms are often used when constructing ground slabs, roadways, driveways, paths etc. (see Figure 18). These are available in straight interlocking lengths that are

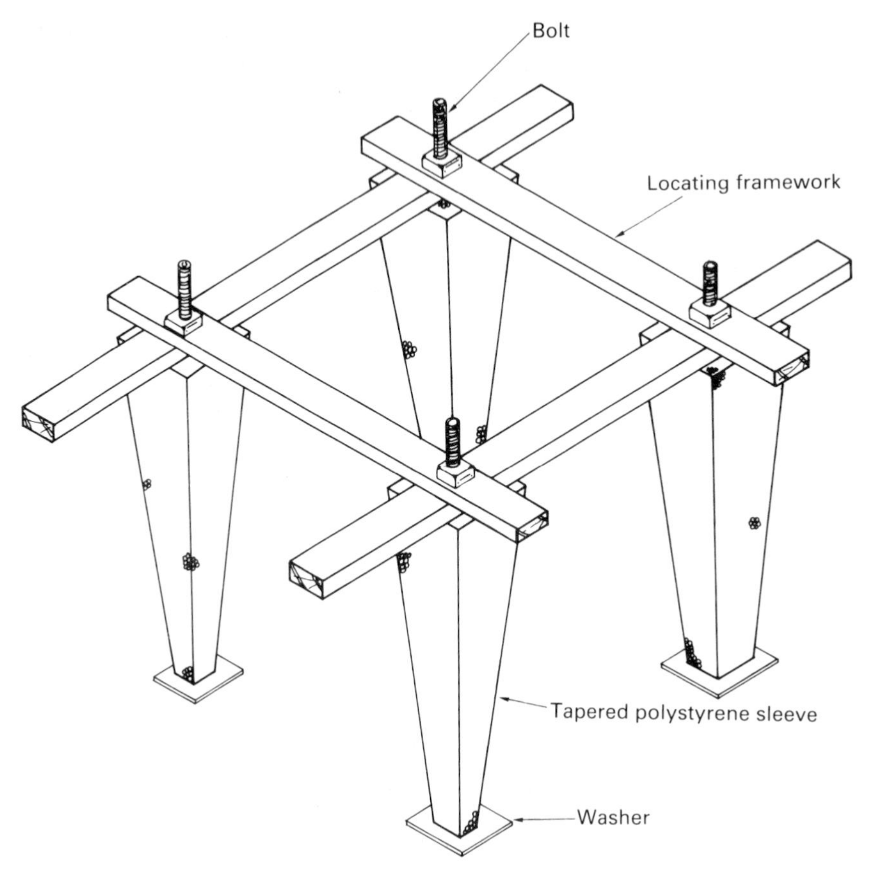

Figure 17 *Locating hold down bolts*

simply staked into the ground and secured by wedging. Flexible types are available where curves are required.

Kickers

Kickers are small upstands of concrete which are used to accurately locate wall and column forms. They are cast on the base concrete between 75 mm and 150 mm in height to the same plan shape as the finished wall or column. Any shallower than this and the kickers' concrete can not be compacted properly. Also they may not bond to the base concrete effectively. Both of these can result in the kicker lifting as the formwork is later tightened around it. Taller kickers, if not exactly vertical, can create problems when plumbing the subsequent formwork.

Although kickers are normally formed as a separate operation after the casting of the base concrete, it is possible to form kickers integrally with a floor slab or foundation.

Figure 19 shows details of typical kicker boxes for walls and columns. The ends are located by housing into the sides. The back of the housing and the edge of the end are splayed; this ensures a tight joint inside the box as it is tightened. Bolts are used to tighten and hold the sides of the kicker boxes together. For easy removal the bolt holes are slotted; a half-turn or so with a spanner when striking is all that is required to remove the bolt. Kicker boxes are normally fixed in position by tying them off the steel reinforcement starter

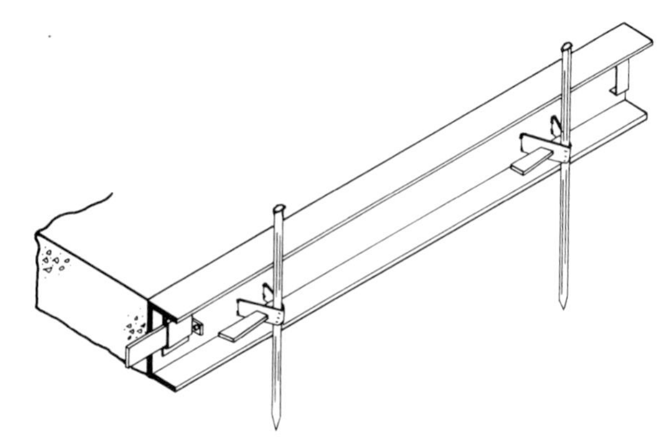

Figure 18 *Road edge form*

bars using notched battens. These battens also act as distance pieces for wall kickers, keeping their sides parallel. In addition, depending on the material used, wall kickers may require additional support along their length to prevent their sides bulging. This can take the form of offcuts nailed into the 'green' concrete.

Kickers that are cast integrally with their base concrete are known variously as floating kickers, nibs or upstands. Invariably they are more difficult to form than by casting separately.

Figure 20 shows typical details of two methods used for forming floating kickers depending on their height. Low kicker edges may be simply supported on previously cast concrete blocks the same thickness as the slab, and tied back to the edge form at intervals using distance pieces.

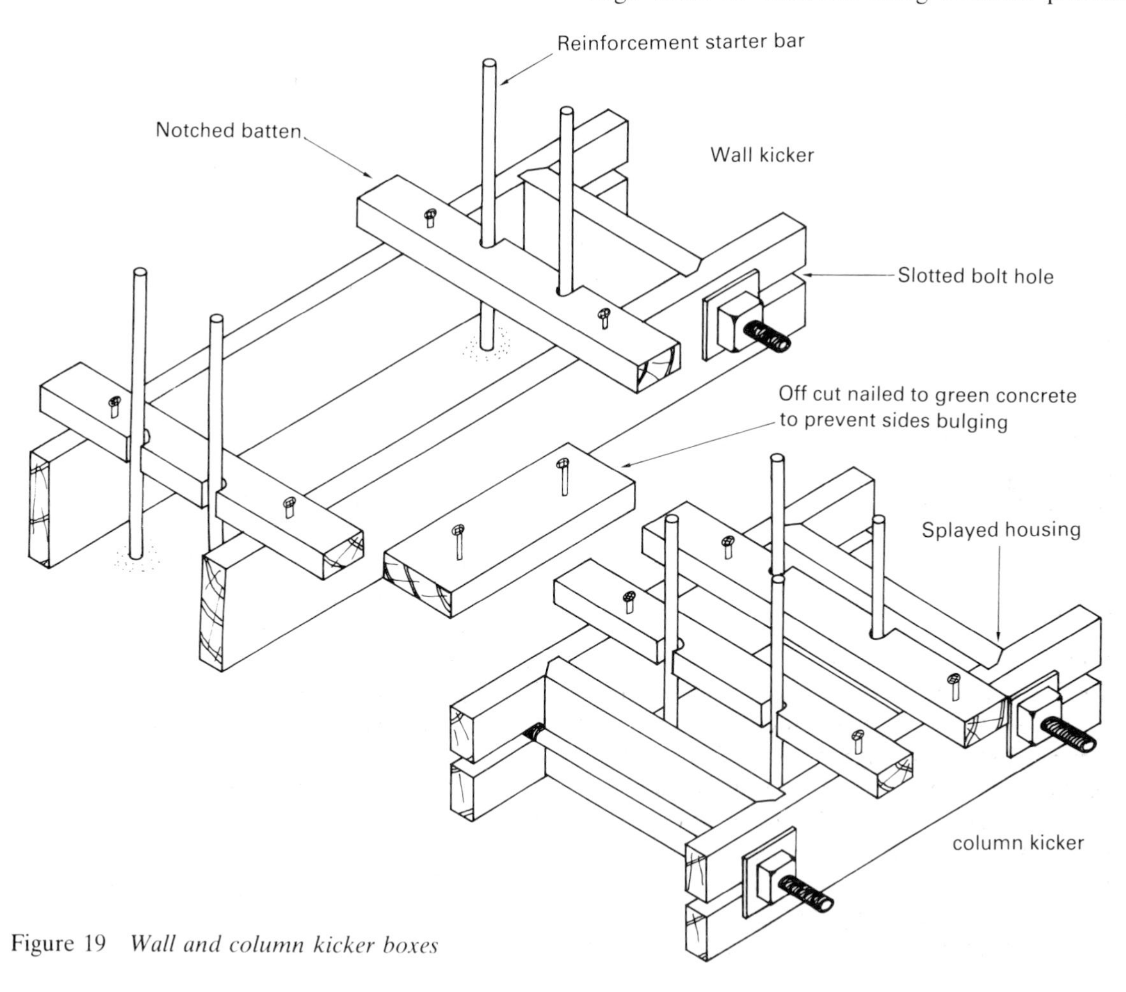

Figure 19 *Wall and column kicker boxes*

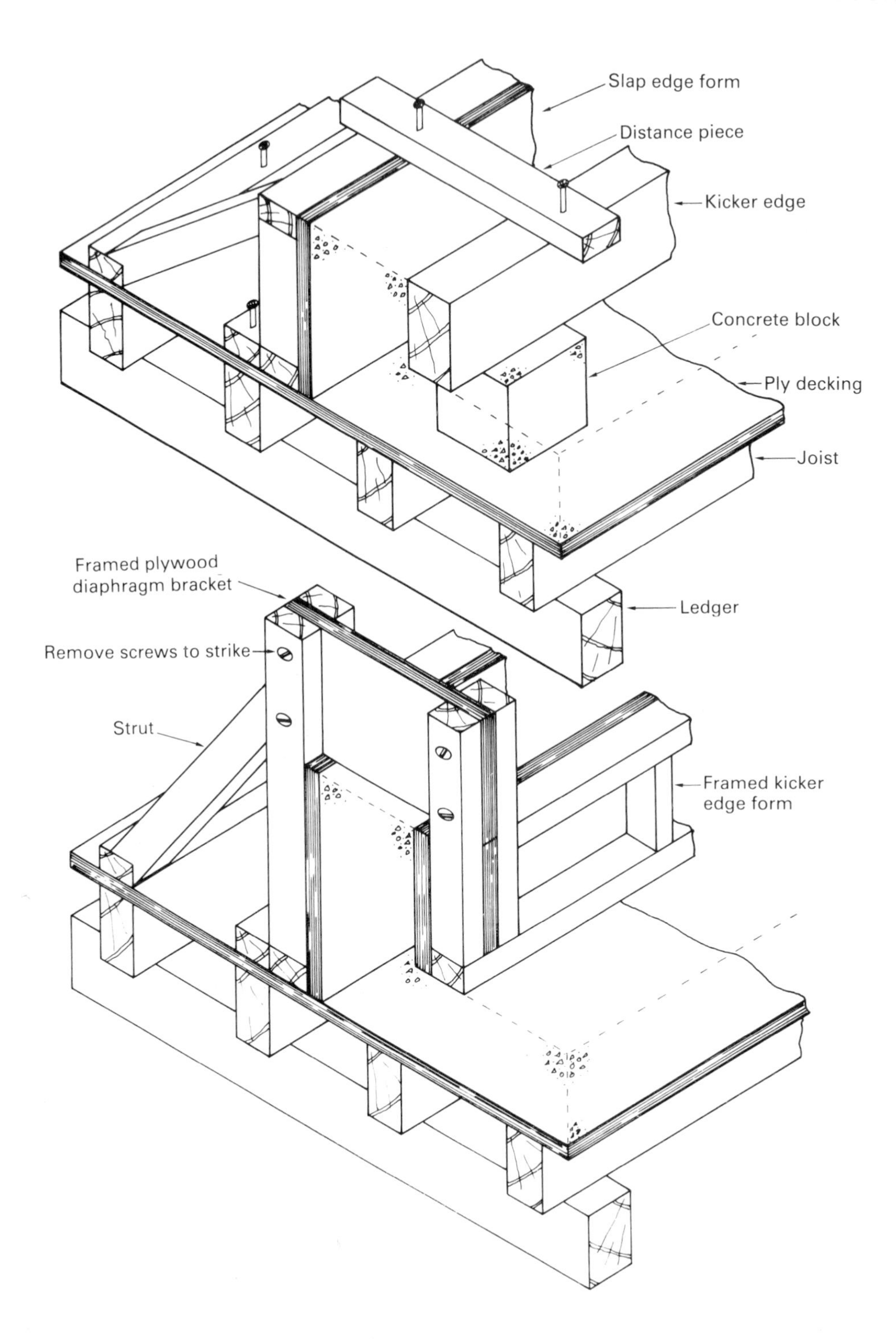

Figure 20 *Floating kickers*

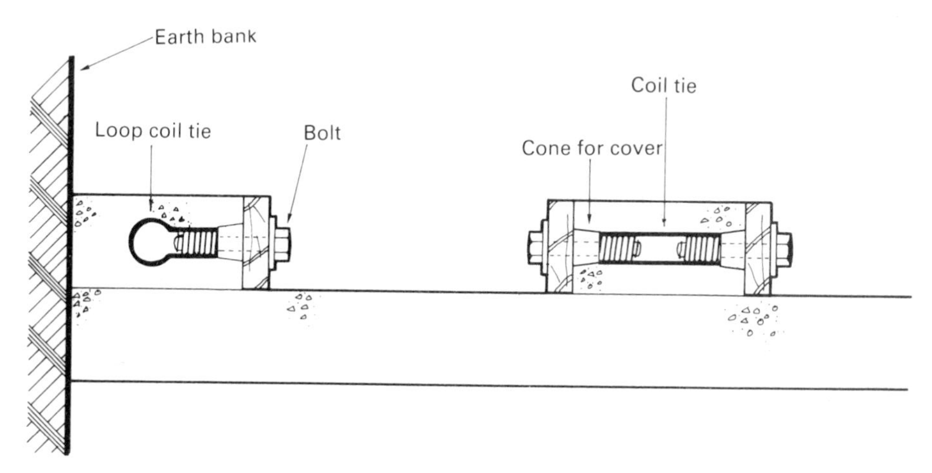

Figure 21 *Kickers with provision for subsequent fixings*

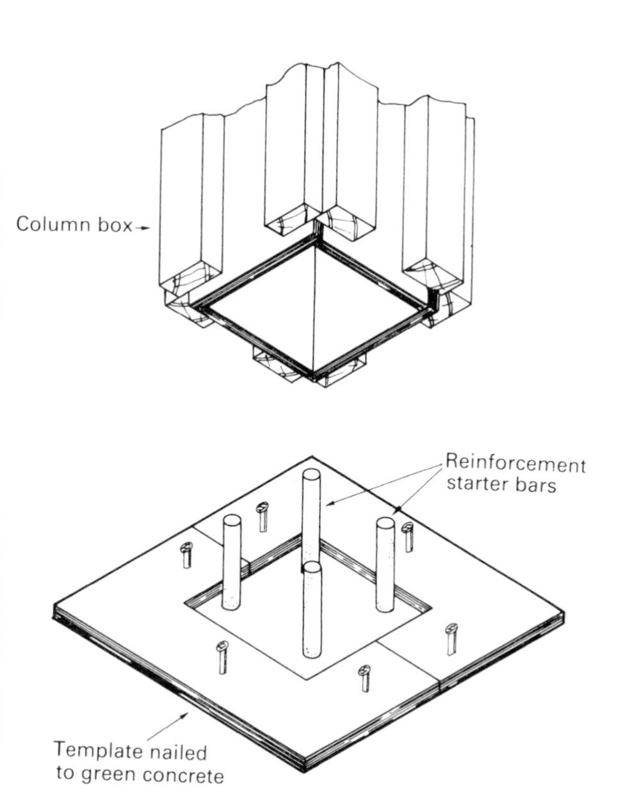

Figure 22 *Use of plywood template*

Deeper kickers and upstand beams may be formed using plywood diaphragm brackets suspended from the slab edge to support the framed kicker edge form. In order to reduce any tendency to sag, concrete blocks may again be used to provide support at intervals.

In certain situations provision in the kicker must be made for the later securing of formwork. This normally takes the form of a cast-in form tie. Figure 21 illustrates a coil tie being cast in a wall kicker. As well as providing a later securing point it ties the kicker sides together, keeping them parallel. Also shown in Figure 21 is a loop coil tie cast in a kicker, where a later anchorage point is only required on one side.

Figure 22 shows a use of plywood templates to position column forms without the need for kickers. This is mainly done for increased speed of production. Although they locate the forms they do not prevent grout loss. The templates themselves are fixed by nailing directly into the green concrete, or may be shot fired using a cartridge fixing tool.

Box-outs and cast-in fittings

Box-outs are indents, recesses, pockets, openings or holes formed in the concrete. There are various methods which can be used in their formation.

Figure 23 illustrates a number of timber box-outs. Their sides have been arranged so that their separate faces can be folded out after casting into the space formed. This method of forming box-outs is sometimes referred to as boxes with chasing corners.

Figure 24 shows a box-out suitable for forming door and window openings in wall forms. This is located between the two form faces of the wall and allows the formwork on both sides to be continuous. Keystone shaped sections are included in each edge of the box. On striking these sections are simply unscrewed, allowing the corners to be removed separately.

Circular or curved box-outs can be formed using plywood fixed around shaped ribs, as shown in Figure 25. To assist removal it has been made in three sections, joined using dovetail

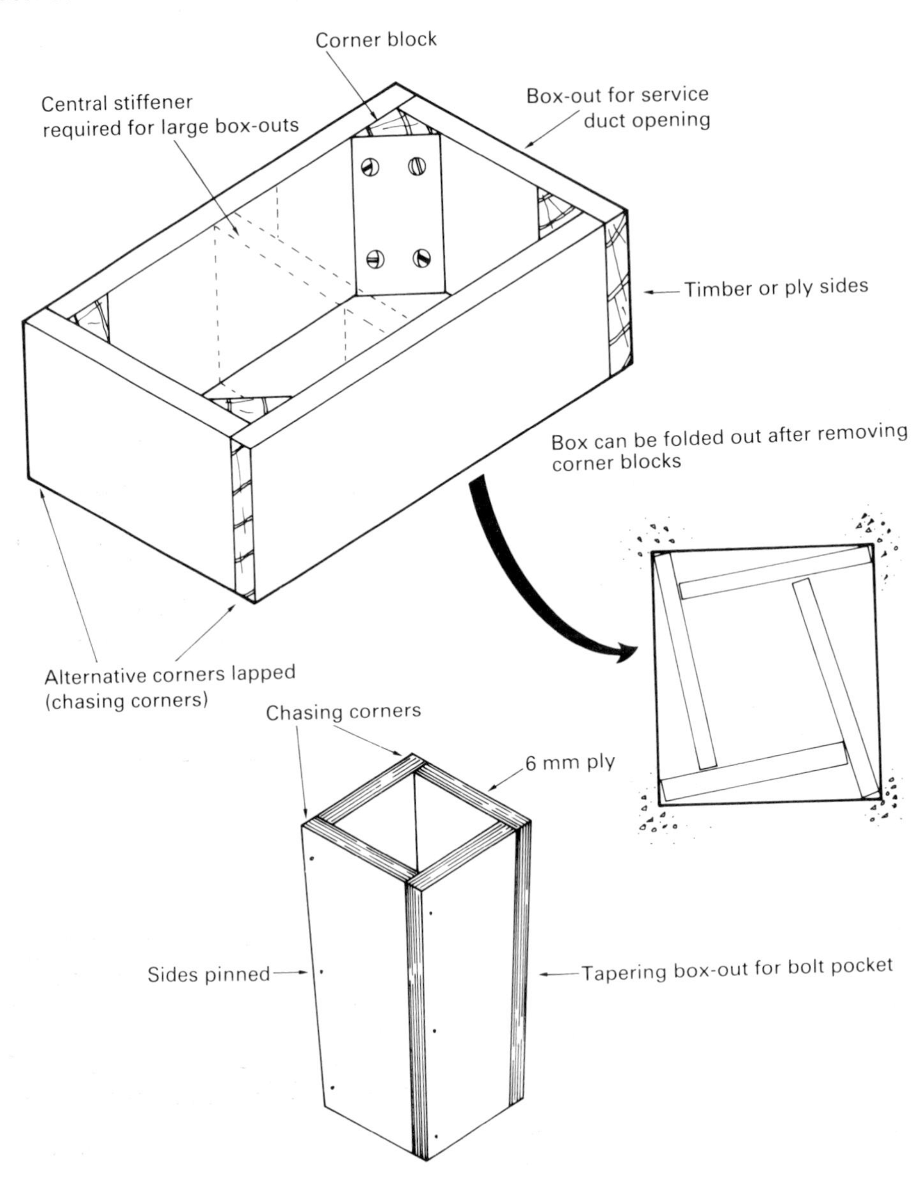

Figure 23 *Box-outs*

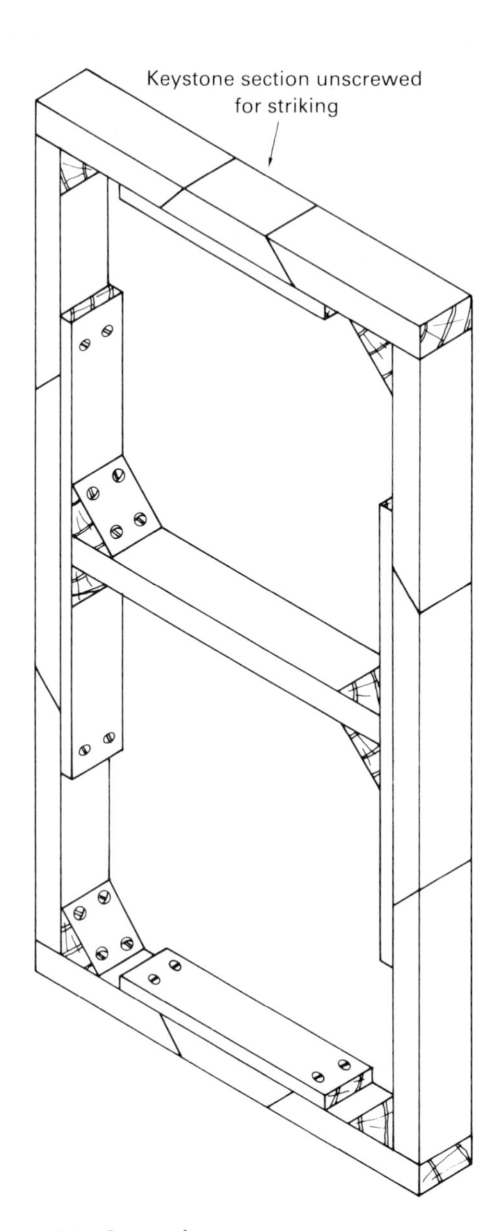

Figure 24 *Square box-out*

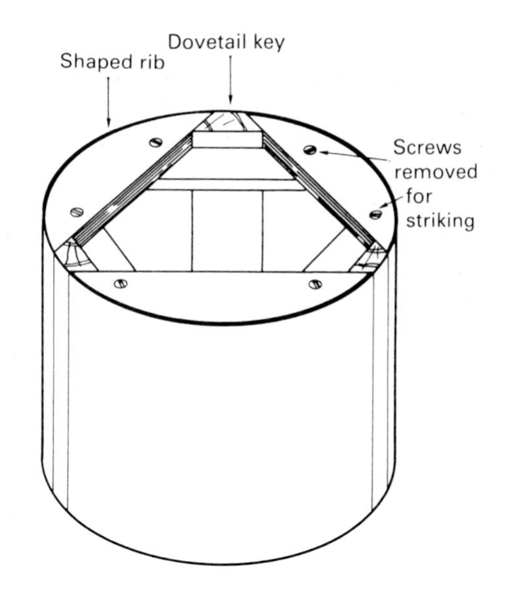

Figure 25 *Circular box-out*

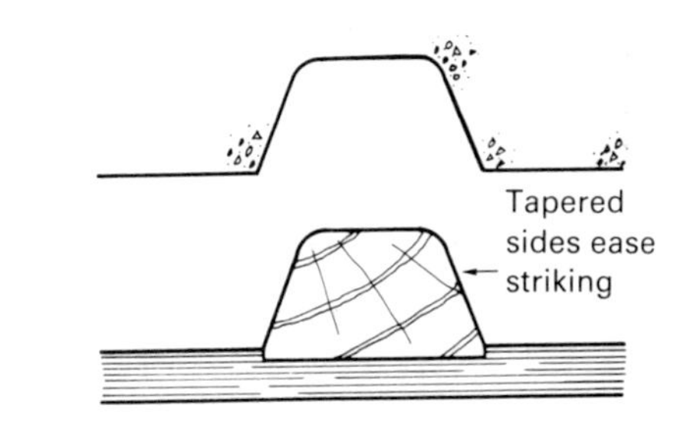

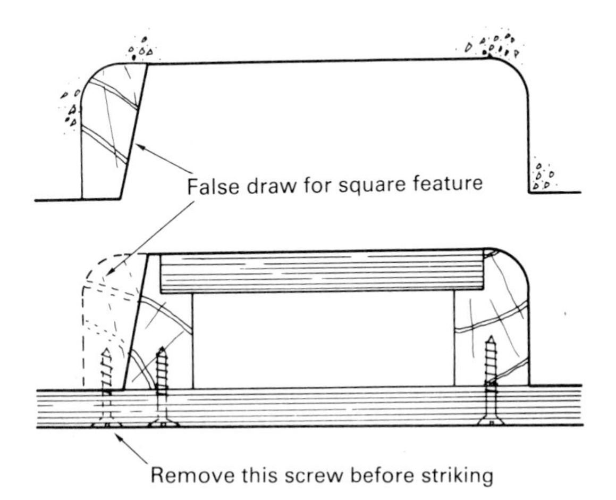

Figure 26 *Draw*

keys. Again on striking the keys are simply unscrewed to enable the sections to be withdrawn with ease into the hole formed.

Where indents or features are required in a concrete face, some form of lead or draw will be required to facilitate striking. Where the detail permits this may be obtained by tapering the sides of the fillet forming the detail. Large indents, or those where the design dictates a square edge, will require some form of false lead or draw. Figure 26 illustrates the method of providing a false draw when forming a rectangular indent with square edges. On striking, the screws through the back of the form face securing the false draw are removed, thus allowing the major part of the box-out to be struck integrally with the main form. After this the false draw can be simply removed and rescrewed back in place ready for any subsequent recasting.

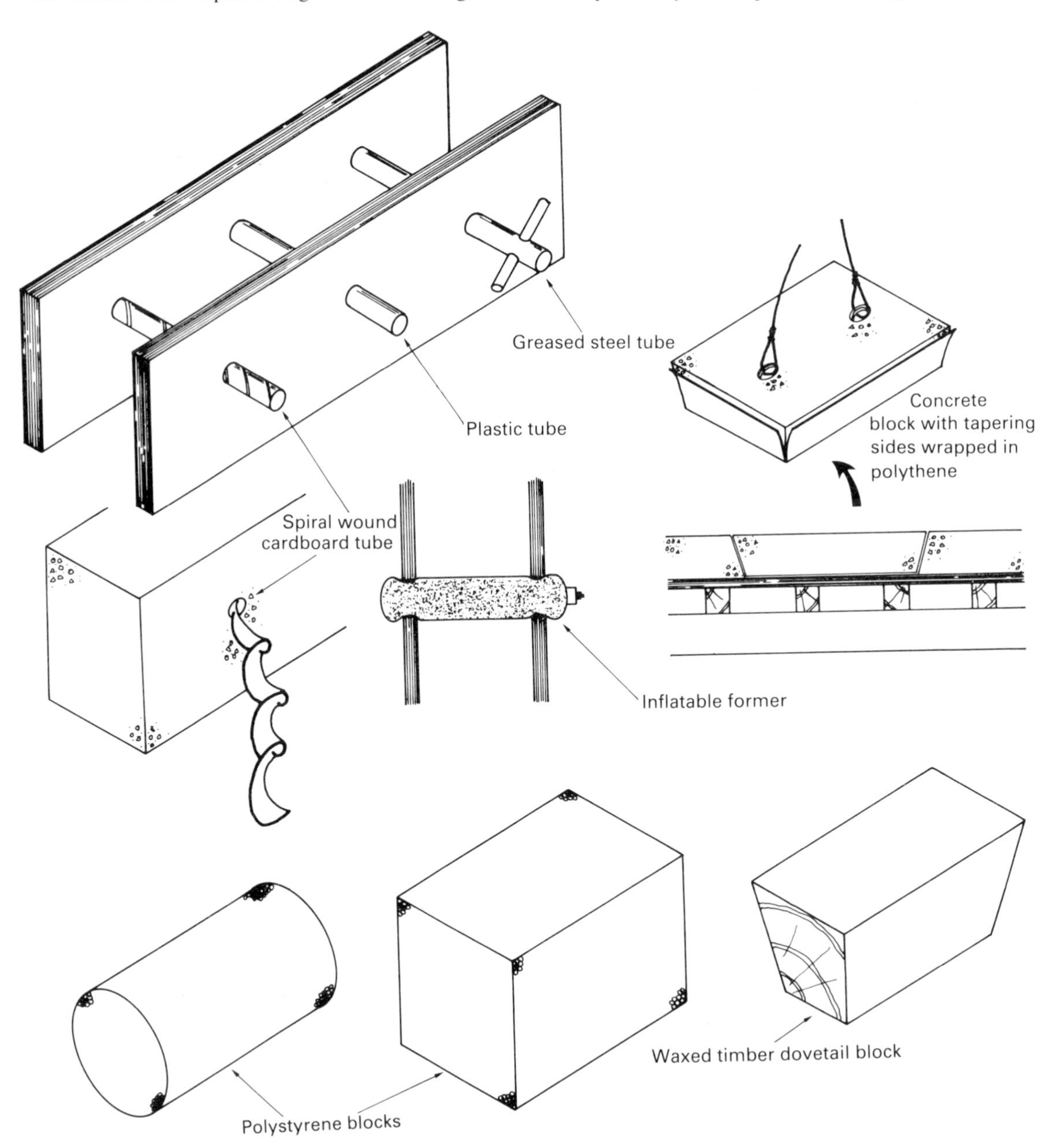

Figure 27 *Methods of forming pockets and holes*

Various other methods of forming pockets and holes are shown in Figure 27:

1 Timber blocks may be dovetailed on four faces and waxed to facilitate removal.
2 Concrete formers are available with tapering sides and wrapped in either thin polystyrene or polythene sheeting. This is an inexpensive method of forming service duct openings etc. in floor slabs. Loops may be cast-in to provide lifting points for later removal by either hand or crane.
3 Spiral wound cardboard tubes, plastic tubes and greased steel tubes are all suitable for forming small diameter holes. On striking the end of the spiral card is simply pulled to unwind it from the hole. Plastic tubes can remain permanently in place, whereas greased steel tubes should be withdrawn from the forms about 2 hours after casting.
4 Inflatable formers are useful for larger diameter holes. They provide grout tight joints at the form faces and are simply deflated for withdrawal.
5 Polystyrene slabs and blocks, being easily cut to shape, are useful for forming a variety of box-outs, although they are only suitable for one-off use and therefore fairly expensive. They also require weighting down to avoid flotation during casting.

Figure 28 illustrates a range of cast-in fittings. These include threaded sockets for later bolt fixings; slotted or dovetail strips for later bolt fixings or brick/masonry ties; and column corner guards for use in car parks and warehouse construction. These are transfer cast into the concrete. They are normally fixed within the formwork prior to casting by bolting into either a threaded portion of the fitting or a captive nut. Before striking the form the bolts are removed, leaving the fitting firmly embedded in the concrete. Self-adhesive strips of foam rubber or Neoprene draught seal can be applied between the form and fitting to prevent grout leakage under or into the fitting. Many short masonry slots are filled with polystyrene, which has to be broken out after casting before the dovetail tongue or tie can be inserted.

Other cast-in fittings that may be used are mainly facilities for services, e.g. electrical conduits, switch boxes, power boxes and other

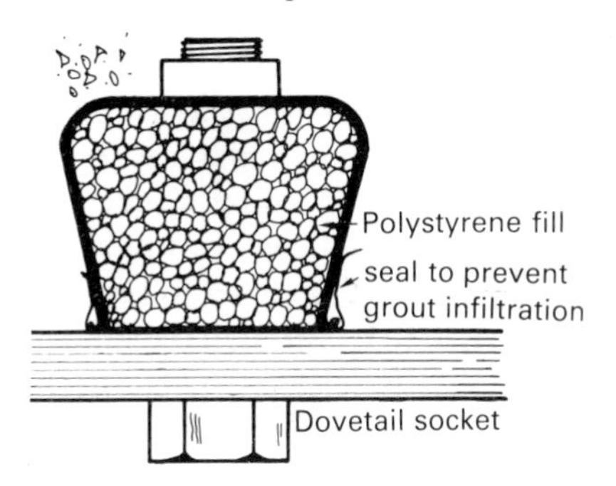

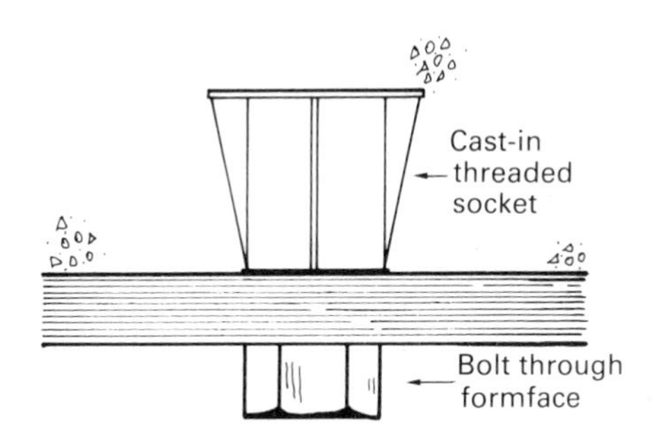

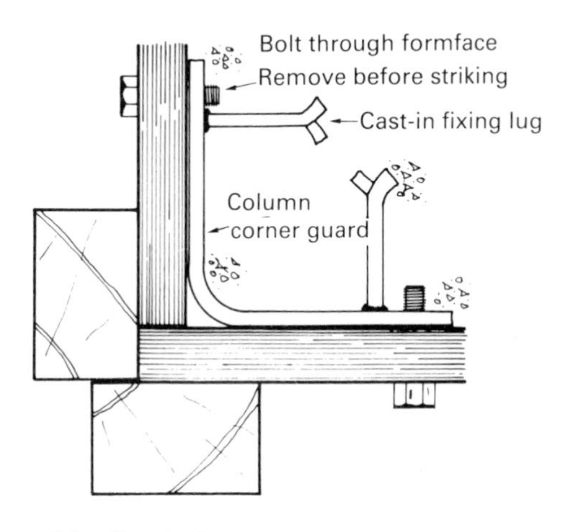

Figure 28 *Cast in fittings*

outlets. These are fixed to the formwork ready for transfer casting, normally by bolting through the form into the fitting. Boxes should be filled with polystyrene and sealed to the forms with foam strip or tape to prevent grout infiltration. Any open conduit ends should be either capped or taped to prevent the possibility of blockage.

Dovetail shaped timber or plastic blocks are often used to provide fixing points for the latter nail or screw fixing of components. These are normally fixed to the forms by either nailing or screwing. Although it is easier as far as positioning is concerned to fix through the block into the form face, rather than through the form face into the block, the projection of the nail or screw from the concrete does cause a safety hazard on striking. Thus where this method is used the formworker must be instructed to hammer over the projection on striking the forms (see Figure 29).

Careful fixing and accurate positioning of all cast-in fittings is essential. The extra time taken on this is minimal compared with the additional costs involved for remedial work owing to missing fittings or inaccuracies in positioning fittings.

Columns

It is common practice to cast columns in one pour up to the level of the underside of the beam or the floor slab.

Figure 30 shows a plan view of a typical square column box which is constructed of four separate timber and plywood sides. Two of the sides are equal to the width of the column and the other two are twice the thickness of the ply oversize. Each side section is made up in such a way that the backing timbers overhang the edge of the plywood by just under the plywood thickness. This corner detail means that the sides will interlock correctly, and makes the column box self-aligning when clamped up.

In order to prevent grout leakage at corners it is essential that the meeting edges of the plywood are straight and square. As an additional precaution Neoprene or foamed rubber self-adhesive strips can be applied between the mating surfaces.

Large column forms may require additional

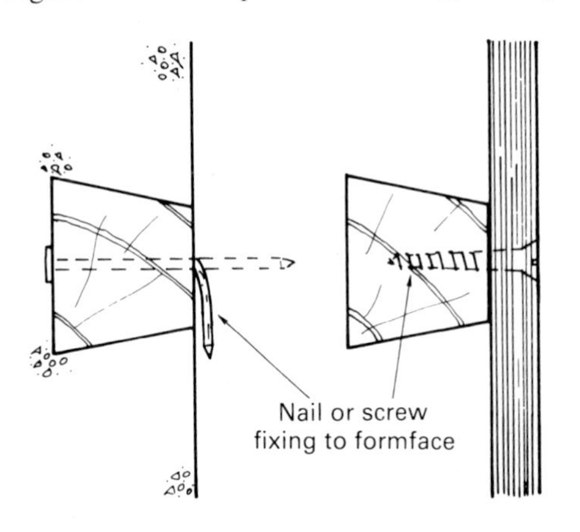

Figure 29 *Fixing blocks*

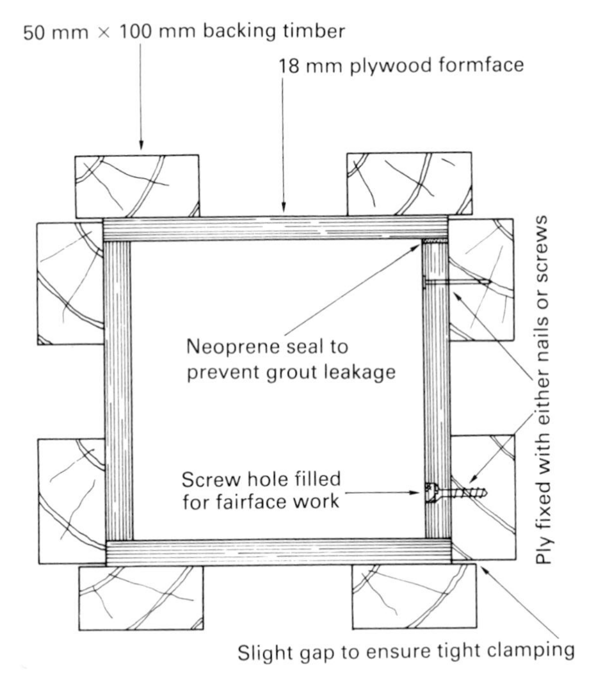

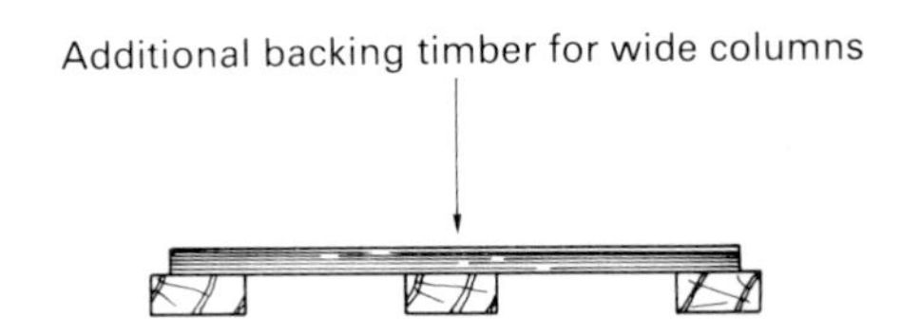

Figure 30 *Square column box plan*

backing timbers on each side or noggins placed behind each clamp between the backing timbers in order to stiffen the plywood and prevent excessive bending.

Where bevelled or bull-nosed corners are required, these can be formed by the use of a fillet. Figure 31 shows the correct method of using fillets which avoids feather edges. If feather edges are used they will result in a poor finish, as on casting they rapidly absorb moisture, curl up and become embedded into the concrete as shown.

Where features are required in column sides, these are best formed by fillets which are either grooved into the plywood face or fitted between separate plywood strips as shown in Figure 32. Again this is to prevent feather edges curling up and becoming embedded.

Figure 33 shows a typical square column box which is assembled in position ready for casting. The four separate side sections are held together with proprietary steel column clamps, the spacings of which are dependent on the design hydrostatic pressure; they are close together at the bottom, further apart toward the top. The whole column assembly is plumbed up and held securely in position by an adjustable steel prop on each side.

An alternative to the use of proprietary column clamps which is particularly useful for very large column boxes is shown in Figure 34. This involves the use of yokes made from walings and tie bolts. The spacings of the yokes in common with proprietary clamps will be dependent on the design hydrostatic pressure.

Timber yokes and bolts are also particularly suited for use where L shaped or rebated columns are being constructed, as illustrated in Figure 35.

After assembly, column forms must be plumbed up in both directions. This can be done with the aid of a suspended plumb bob or a

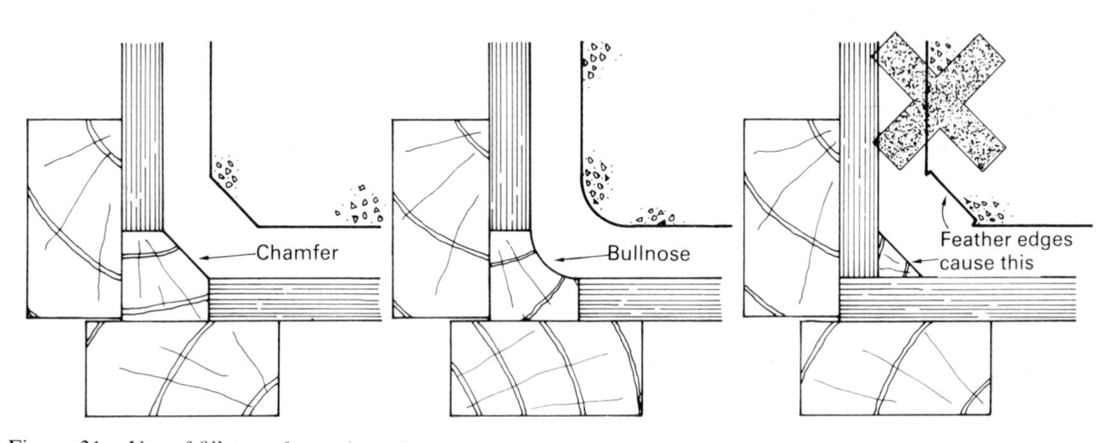

Figure 31 *Use of fillets to form shaped corners*

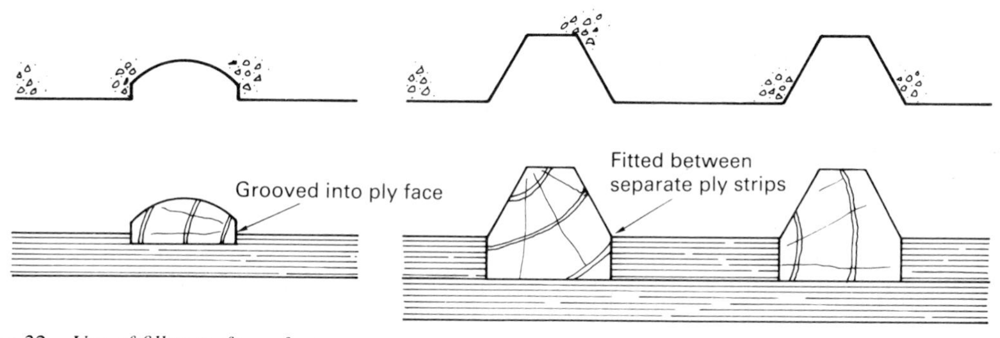

Figure 32 *Use of fillets to form features*

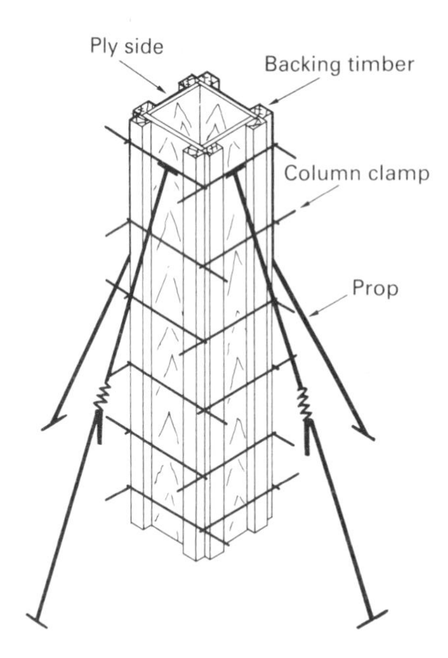

Figure 33 *Column ready for casting*

straight edge and level. Figure 36 illustrates both of these methods. Opposite props are adjusted until the column is plumb in that direction. Ensure that when tightening any prop the opposite one is slackened by an assistant at the same time. Failure to do this will result in upward pressure, causing the whole column box to be lifted from the kicker or the kicker from its base. This plumbing operation must be repeated on the adjacent side so that the column is plumb in both directions.

Four standard proprietary steel framed panels, either steel or plywood faced, can be used to form square and rectangular columns. These are connected at their external corners with L shaped angles which are secured with keys and wedges, as illustrated in Figure 37.

Occasionally columns are topped with a splayed or mushroom shaped head. Although they form a distinctive decorative feature, their main purpose is in fact to carry the load of large span floor slabs and transfer these loads on to the columns. The more common method of supporting and transferring loads is to use a system of beams.

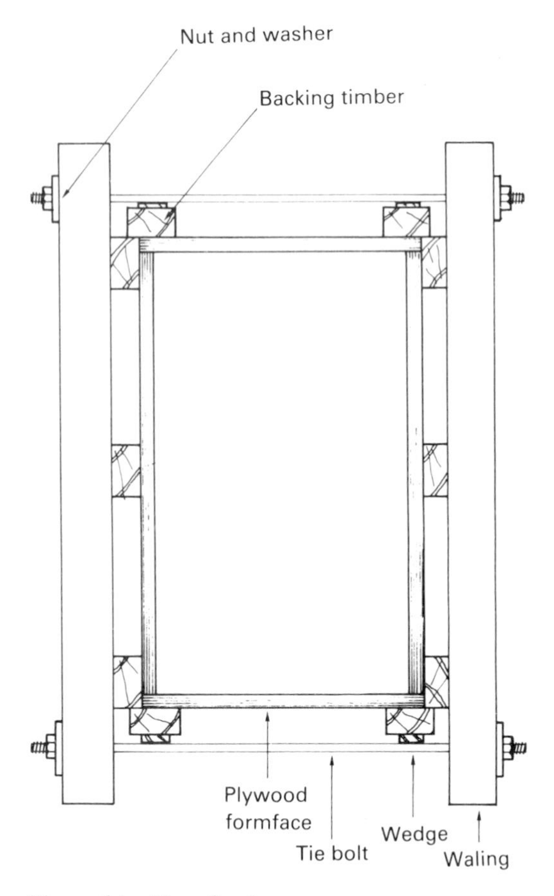

Figure 34 *Use of yokes*

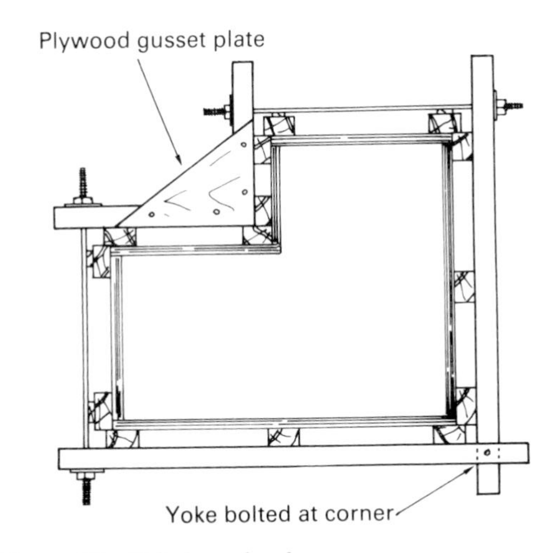

Figure 35 *'L' shaped column*

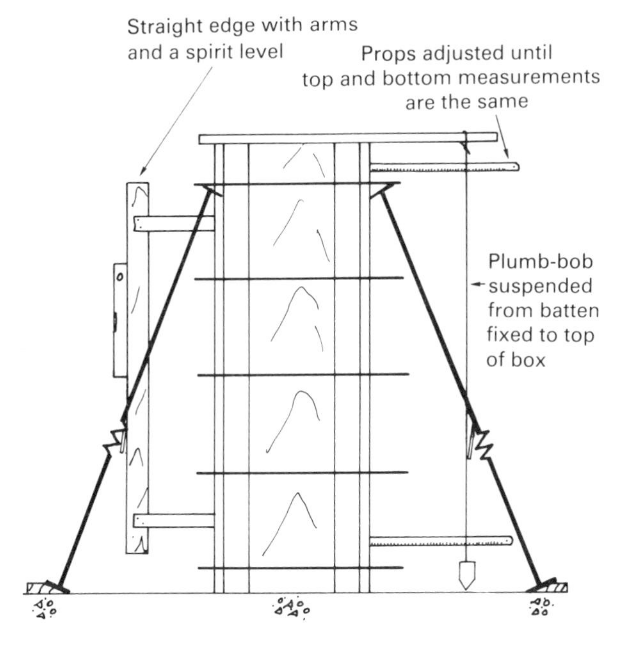

Figure 36 *Plumbing a column box*

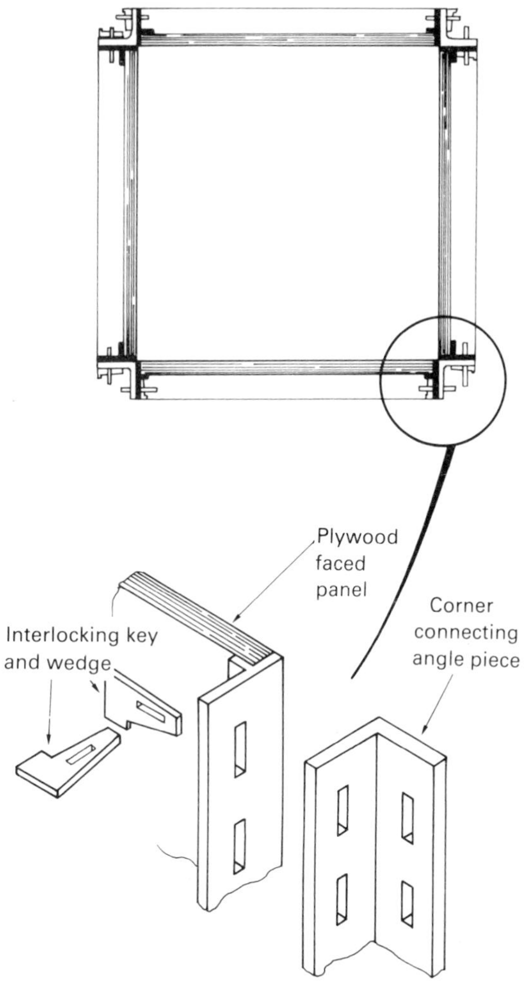

Figure 37 *Proprietary column form*

Where large numbers of mushroom heads are involved, the use of purpose-made steel or glass reinforced plastic forms would be an economical consideration. Illustrated in Figure 38 is a method for forming mushroom heads using timber and plywood. The column is first cast up to the level of the splay. A timber yoke is bolted on to the column shaft at its top to form a platform on to which the head formwork can be erected. This consists of two long plywood sides fixed to shaped plywood formers and bolted together on top of the yoke platform. The other two sides are cut and fixed on to the first two.

Circular columns can be constructed using cardboard tube formers, proprietary glass reinforced plastic formers, proprietary steel formers, or made-up timber and plywood forms.

Figure 39 illustrates a section through a cardboard tube former. These require backing timbers and either yokes or metal banding to prevent distortion from the circular during casting. The example shown uses metal banding of the type used in heavy packaging. The bands are tightened around the forms using a special tightening and crimping device. On striking the bands are simply cut with shears to release the backing timbers. It is customary to leave the cardboard tube in position until occupation in order to provide physical protection for the concrete. Being spirally wound, it is then simply unwound from the column. This method has the advantage of being fairly economical, but the tubes are only suitable for one-off usage.

Figure 40 illustrates a method of forming circular columns in timber and plywood. It is constructed in two halves each consisting of plywood formers, timber laggings and vertical

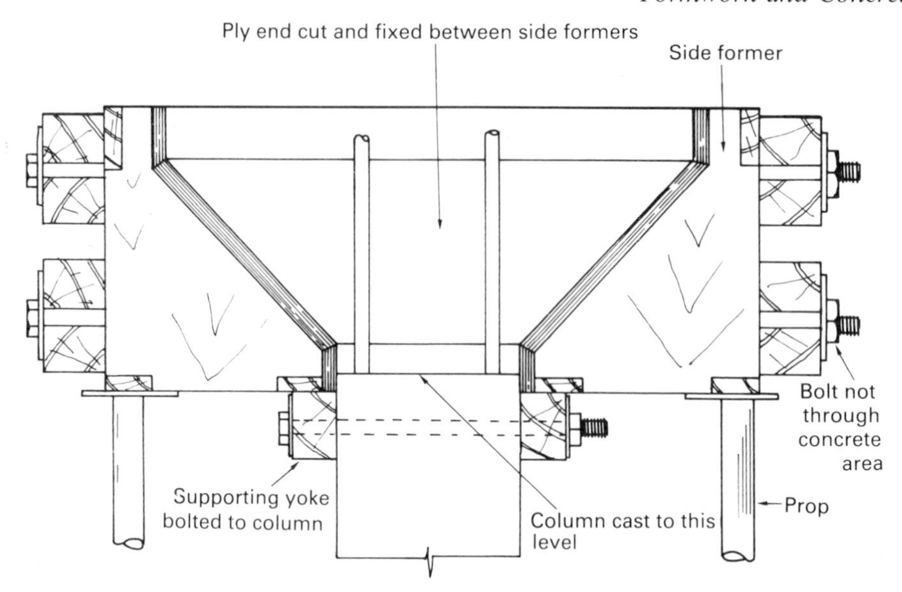

Figure 38 *Plywood mushroom head form*

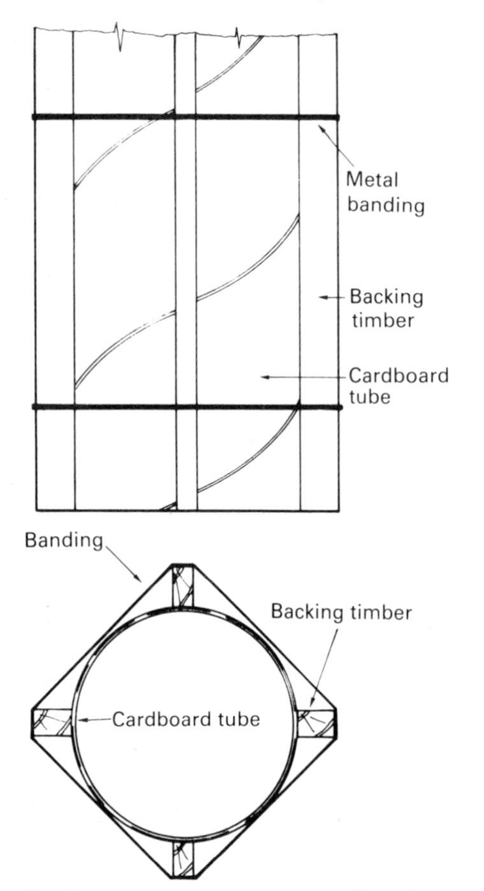

Figure 39 *Circular column using a cardboard tube*

studs. The spacing between the ply formers will be similar to that used for column clamps or yokes. On assembly the two separate halves are bolted together around the kicker.

Figure 41 illustrates an alternative method of constructing circular columns. This uses solid timber yokes and laggings. Where a fair faced finish is required the laggings may be lined with plywood or hardboard. Four pieces of shaped solid timber are used to form each yoke. Two of these are permanently cleated together to form a semi-circle. Subsequently the two halves are bolted together on assembly.

Circular column forms are plumbed up and held in position using adjustable steel props, by the same method as that for square columns.

Proprietary or purpose-made circular column forms in either steel or glass reinforced plastic, although initially very expensive, can be a cost effective consideration where large numbers of columns are required. This is more especially the case in situations where mushroom headed columns are required.

Figure 42 illustrates a section through a column box with shaped ends and a box-out to create a service duct. This is in fact a combination of a square and circular column. It has been made up in two halves using plywood formers, vertical studs and timber laggings lined with either ply-

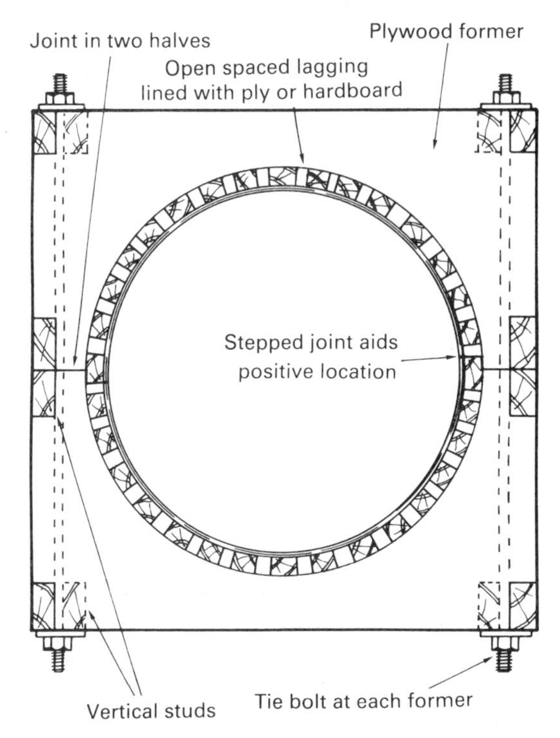

Figure 40 *Circular column using a plywood former*

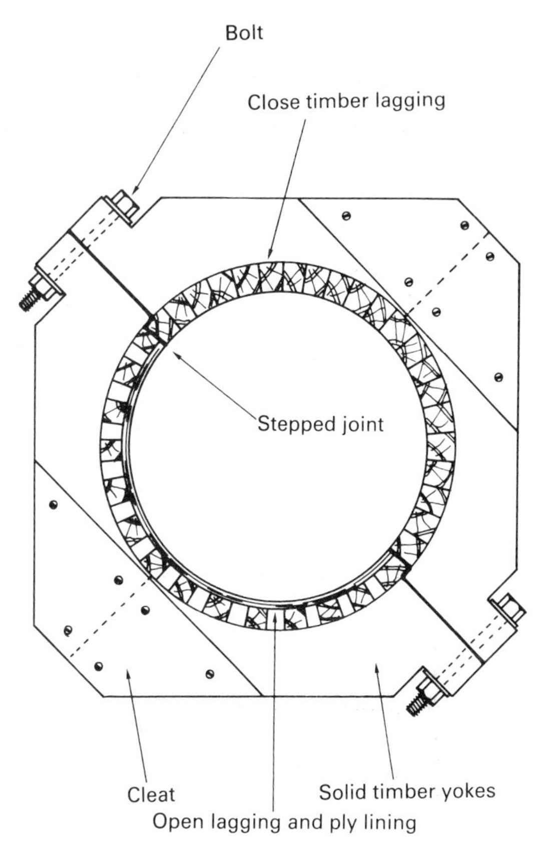

Figure 41 *Circular column using a timber yoke*

wood or hardboard. The rectangular service duct has been formed using a box-out that incorporates a false draw.

The plywood former and laggings method of column box construction is very versatile; it can be adapted to create any shape of column box.

Walls

Wall formwork normally consists of a series of standard wall panels secured together. These may be either proprietary steel framed with ply or steel faces, or timber framed with ply faces. These panels are then tied over their backs with either horizontal walings or vertical soldiers. The decision on whether to use vertical form panels with horizontal walings or horizontal panels and vertical soldiers (Figure 43) depends on the designer's preference and the nature of the work. In general when forming high lifts where considerable hydrostatic pressures are expected, it is

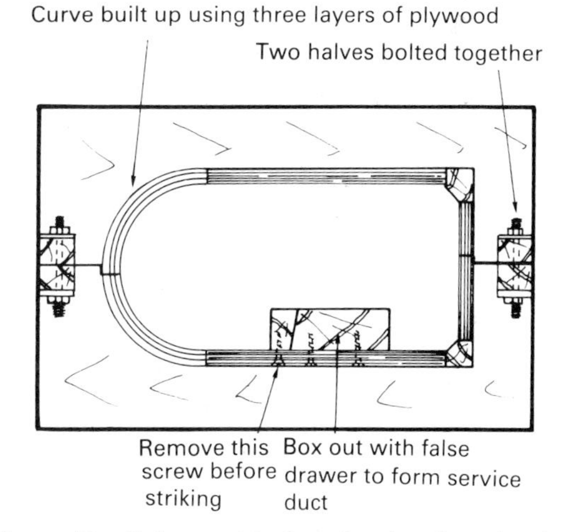

Figure 42 *Column with shaped end and service duct*

common practice to use the panels horizontally and secure them with vertical soldiers.

Illustrated in Figure 44 is a typical wall form, using standard timber-frame ply-faced panels, twin timber soldiers and form ties. These panels are erected against a previously cast kicker. Adjustable steel props are positioned on both sides of the wall at about 1 m centres to provide a means of plumbing the forms and holding them in position. The props must be adequately seated at both ends to prevent slipping. Too steep a prop angle should be avoided as this tends to lift the form panels.

An alternative to adjustable steel props is push-pull props that bolt to the form and the substructure, enabling them to be used on one side of the form only.

Form tape may be used to seal the plywood joints where adjacent panels butt up, in order to prevent grout loss. However, the imprint of the tape may not be acceptable for fair face finishes (see Figure 45).

Proprietary steel soldiers or strongbacks may be used for extra strength where high walls are cast in one lift. Alternatively they may be used to limit the number of form ties required (see Figure 46).

Form ties are used to maintain the spacing between the two form faces and also resist the pressures of fluid concrete. The type used and their spacing will depend on the design P_{max}. Table 17 illustrates the main range of ties and for each states typical uses.

Walls are often cast in relatively short sections to allow contraction to take place before adjacent sections are cast. This normally involves making provision in the stop end for reinforcement to protrude into the adjacent section. The detail illustrated in Figure 47 can be used for this purpose. The ply stop end is cut in a zigzag pattern to accommodate the steel and enable it to be simply slid out from either side on striking.

Careful formwork detailing of corners and attached piers is required in order to resist the increased pressures at these points. Typical details are shown in Figure 48.

Walls may also be formed using proprietary

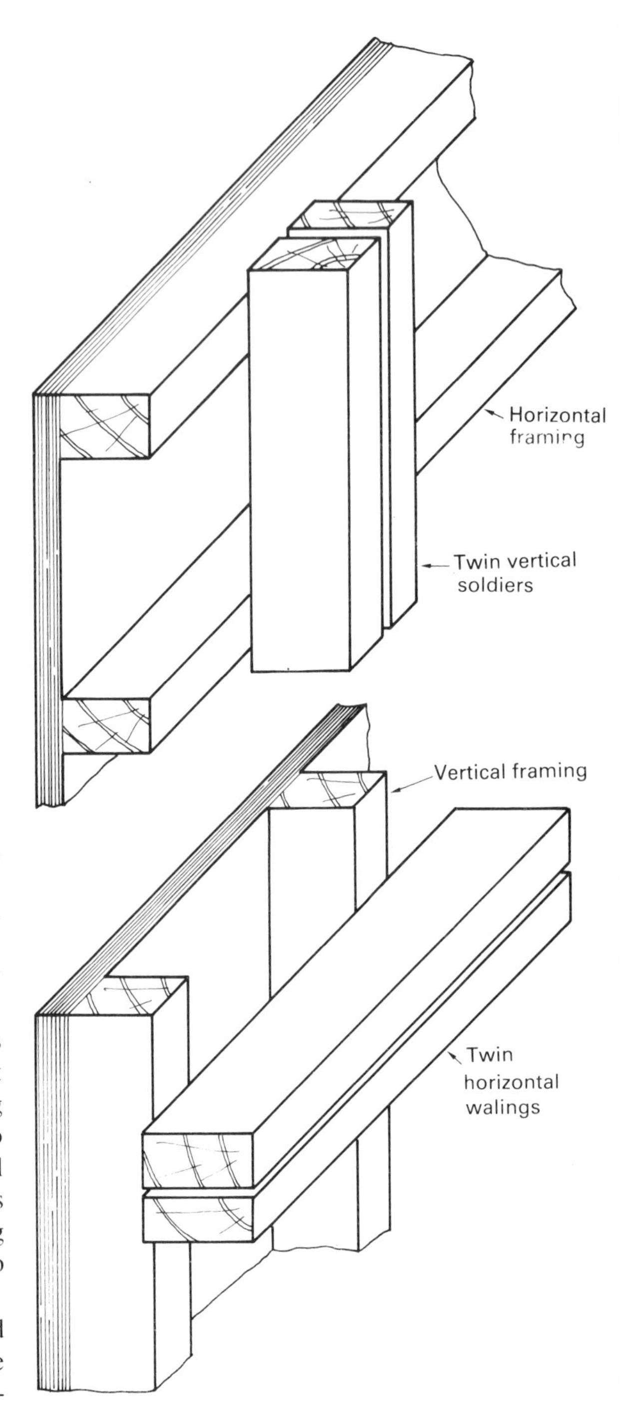

Figure 43 *Wall formwork*

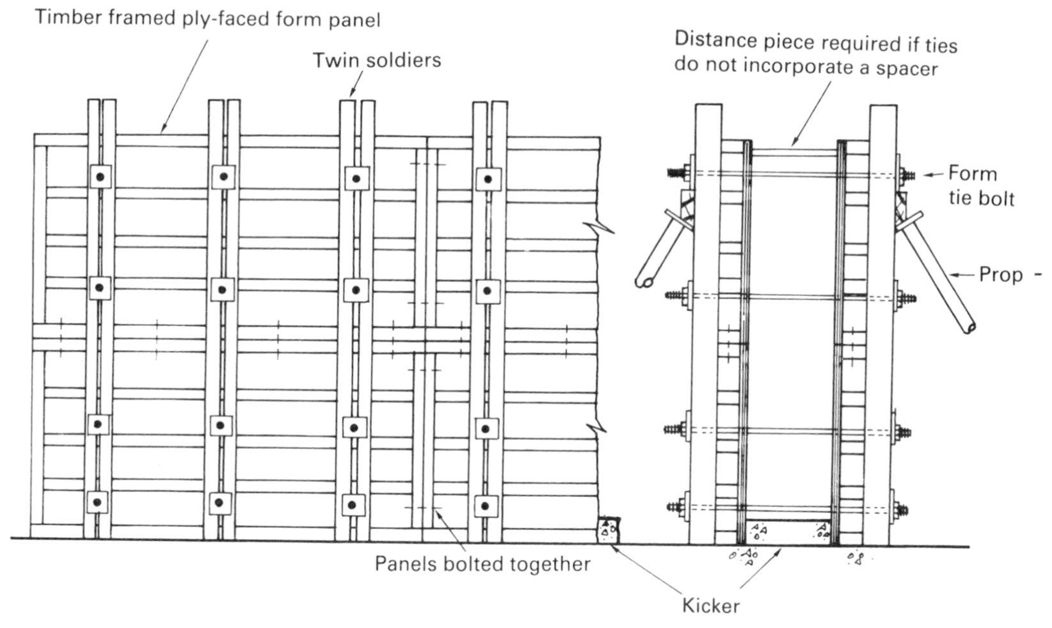

Figure 44 *Wall formwork*

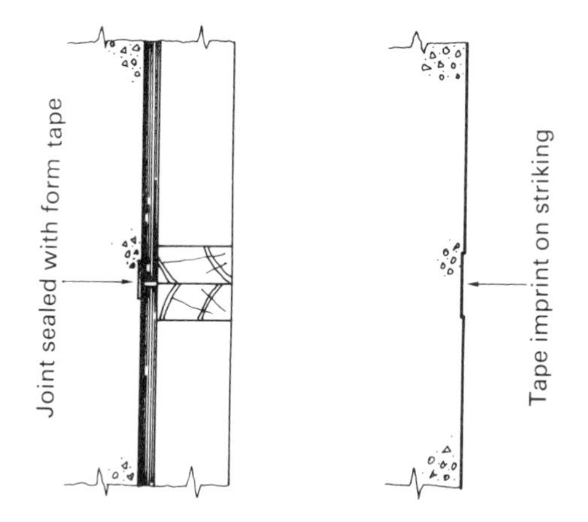

Figure 45 *Use of formtape*

systems. Most of these consist of steel framed panels jointed together using keys and wedges. Figure 49 illustrates details of one such system. Scaffold tubes are used as walings to tie the panels and align the wall.

Walls are sometimes cast in a number of vertical lifts. The main reasons for this are to restrict the development of high pressures that would occur if the wall was cast in one pour, and also to provide an economical formwork detail by using small section material and obtaining the maximum reuse from the minimum number of form panels.

The formwork for walls cast in this way is known as *climbing wall* formwork. Figure 50 shows a basic climbing formwork arrangement. After casting and curing the form panels are struck and raised to form the next lift. The climbing process is repeated as required until the desired wall height is reached.

The joint line between adjacent lifts causes problems owing to difficulty in sealing the upper formwork to the previously cast surface. Often this area is marred by sand textured areas and curtains. This problem can be overcome by making a recessed feature of the joint using fillets attached to the form face, as illustrated by Figure 51. The use of this recessed feature is particularly suitable for vertical sawn board lined forms, where the natural grain variation make it almost impossible to create a neat grout tight joint.

In certain circumstances concrete walls are required to be cast up against earth banks or other structures. In these cases single sided wall forms are required. As it is not possible to obtain

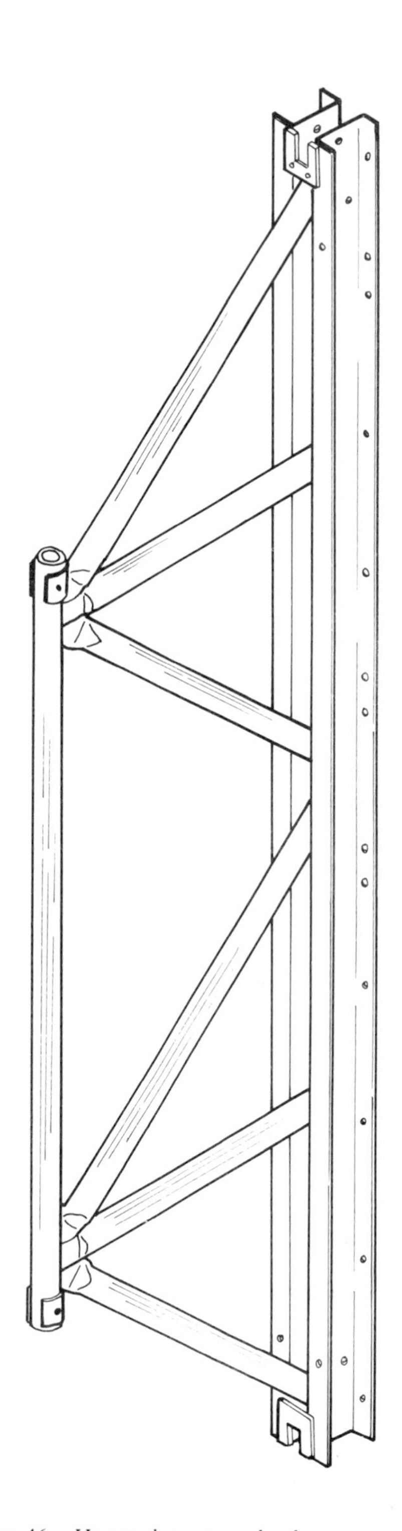

Figure 46 *Heavy duty strongback*

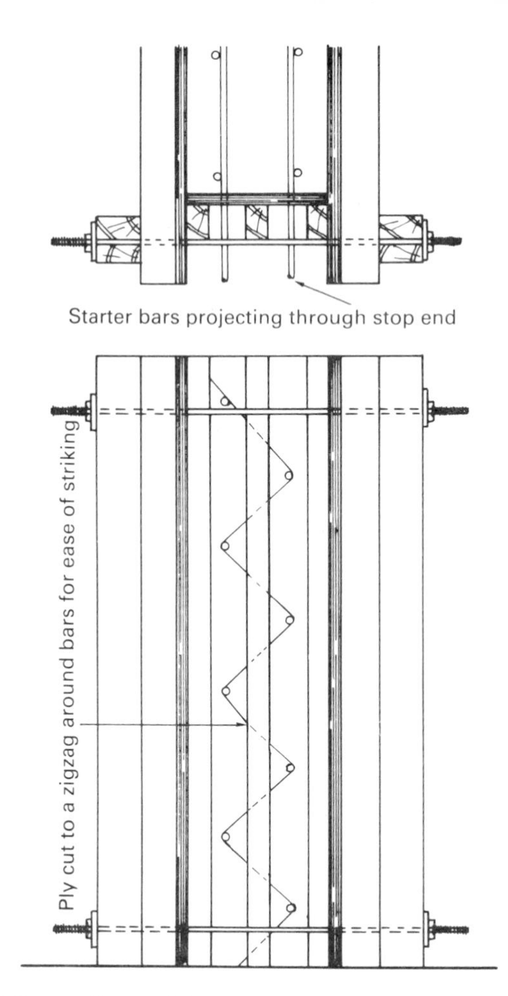

Figure 47 *Wall form stop end*

intermediate support from form ties, a more rigid structure is required. The single sided wall form illustrated in Figure 52 uses standard form panels and heavy duty proprietary strongbacks that incorporate their own jacking and levelling provision, as well as a working platform for concreting operations. To prevent uplift at the base of the forms it is essential to securely tie them down. This is achieved in the detail shown by bolting the strongbacks to loop coil ties which have previously been cast in the kicker.

The construction of curved concrete walls may be carried out using panels made up from shaped horizontal ribs, vertical laggings and plywood

Table 17 Formties

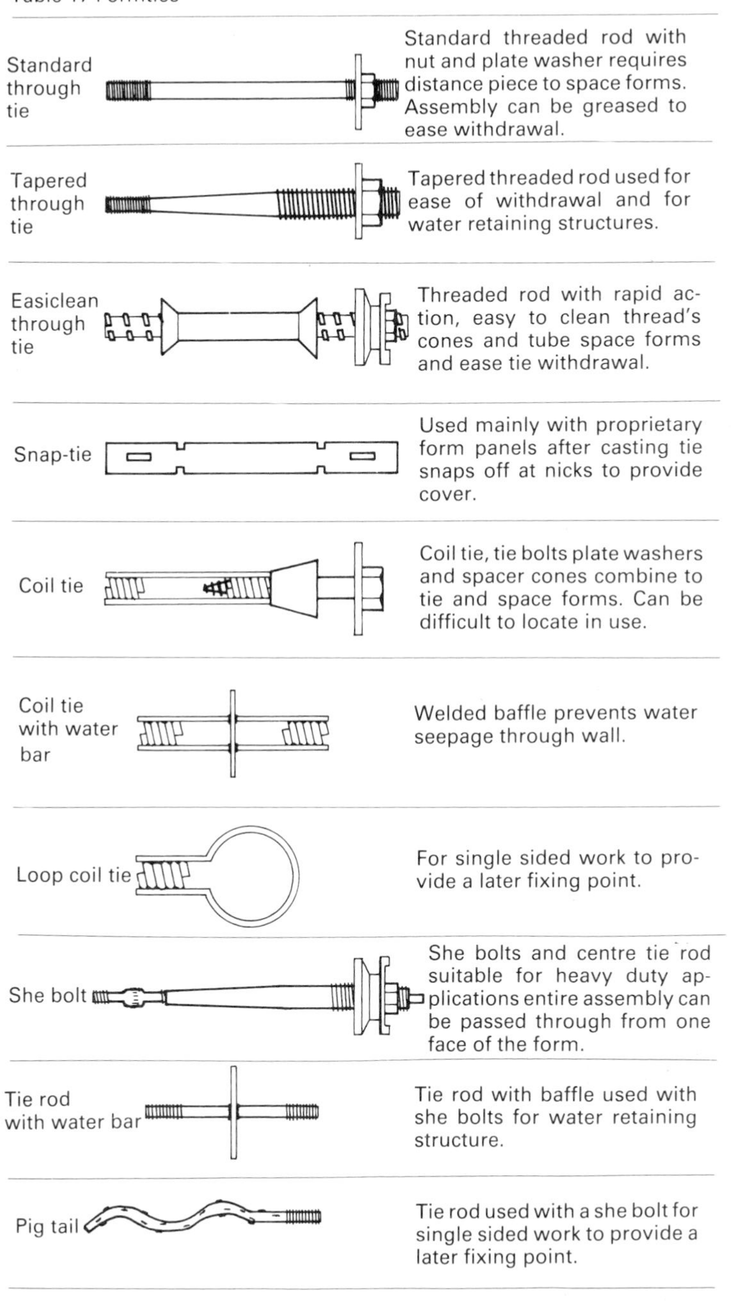

Type	Description
Standard through tie	Standard threaded rod with nut and plate washer requires distance piece to space forms. Assembly can be greased to ease withdrawal.
Tapered through tie	Tapered threaded rod used for ease of withdrawal and for water retaining structures.
Easiclean through tie	Threaded rod with rapid action, easy to clean thread's cones and tube space forms and ease tie withdrawal.
Snap-tie	Used mainly with proprietary form panels after casting tie snaps off at nicks to provide cover.
Coil tie	Coil tie, tie bolts plate washers and spacer cones combine to tie and space forms. Can be difficult to locate in use.
Coil tie with water bar	Welded baffle prevents water seepage through wall.
Loop coil tie	For single sided work to provide a later fixing point.
She bolt	She bolts and centre tie rod suitable for heavy duty applications entire assembly can be passed through from one face of the form.
Tie rod with water bar	Tie rod with baffle used with she bolts for water retaining structure.
Pig tail	Tie rod used with a she bolt for single sided work to provide a later fixing point.

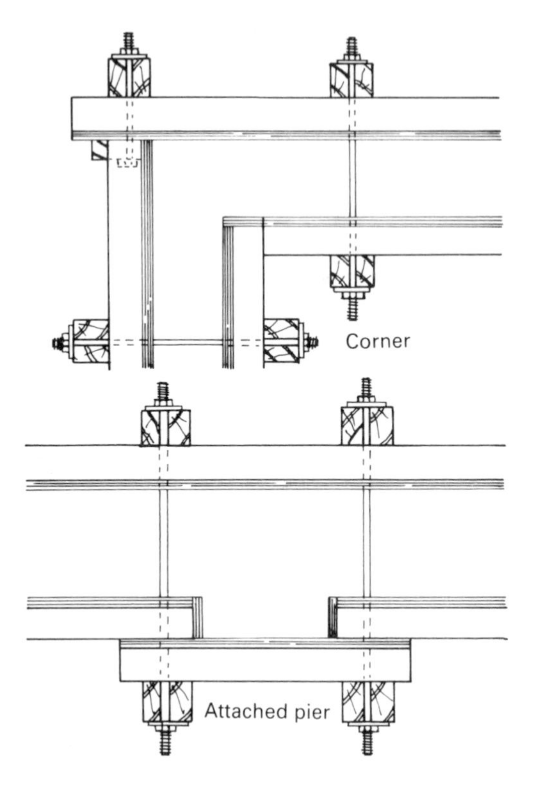

Figure 48 *Wall details*

form face linings, as shown in Figure 53. Two different shaped form panels are required – one set to the concrete face internal radius, and the other to the external radius. Twin vertical soldiers and form ties are used to secure the panels where the ribs overlap.

Illustrated in Figure 54 is a proprietary system for forming curved wall shapes. This utilizes strongbacks attached to the plywood form face. The desired internal and external curvatures are formed by adjusting the turn buckles which connect the edge angles that are attached to both ends of the plywood sheets. Adjacent forms may be joined by bolting through these edge angles.

Slabs and beams

The decking and support structure for floor and roof slabs will normally take the form of either plywood decking on softwood joists supported by softwood ledgers, which are in turn supported on adjustable steel props; or one of the various proprietary formwork systems available; or a partial combination of both, typically a proprietary support system with traditional ledgers, joists and a plywood form face.

Figure 55 illustrates the traditional formwork arrangement for slabs. The spacings of the various formwork members and sectional sizes for the timbers are dependent on the slab thick-

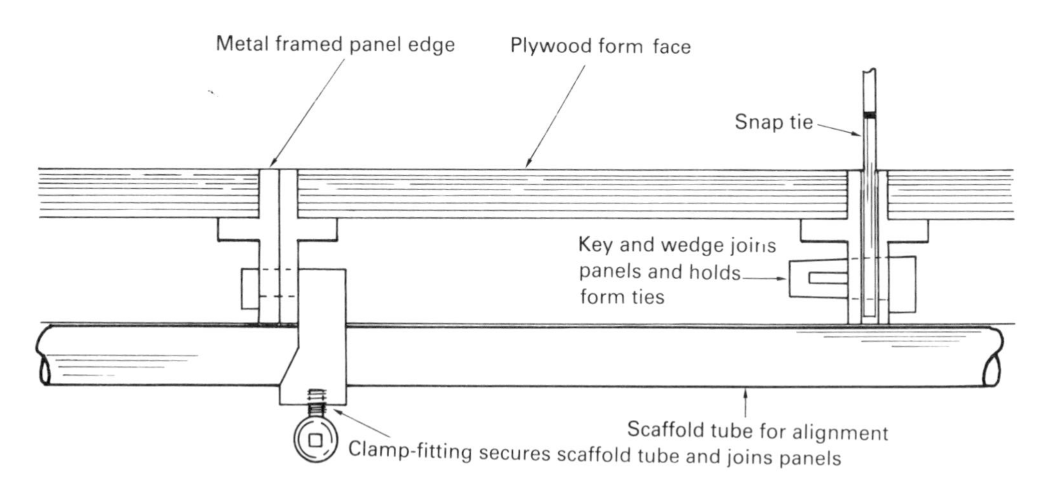

Figure 49 *Proprietary wall form details*

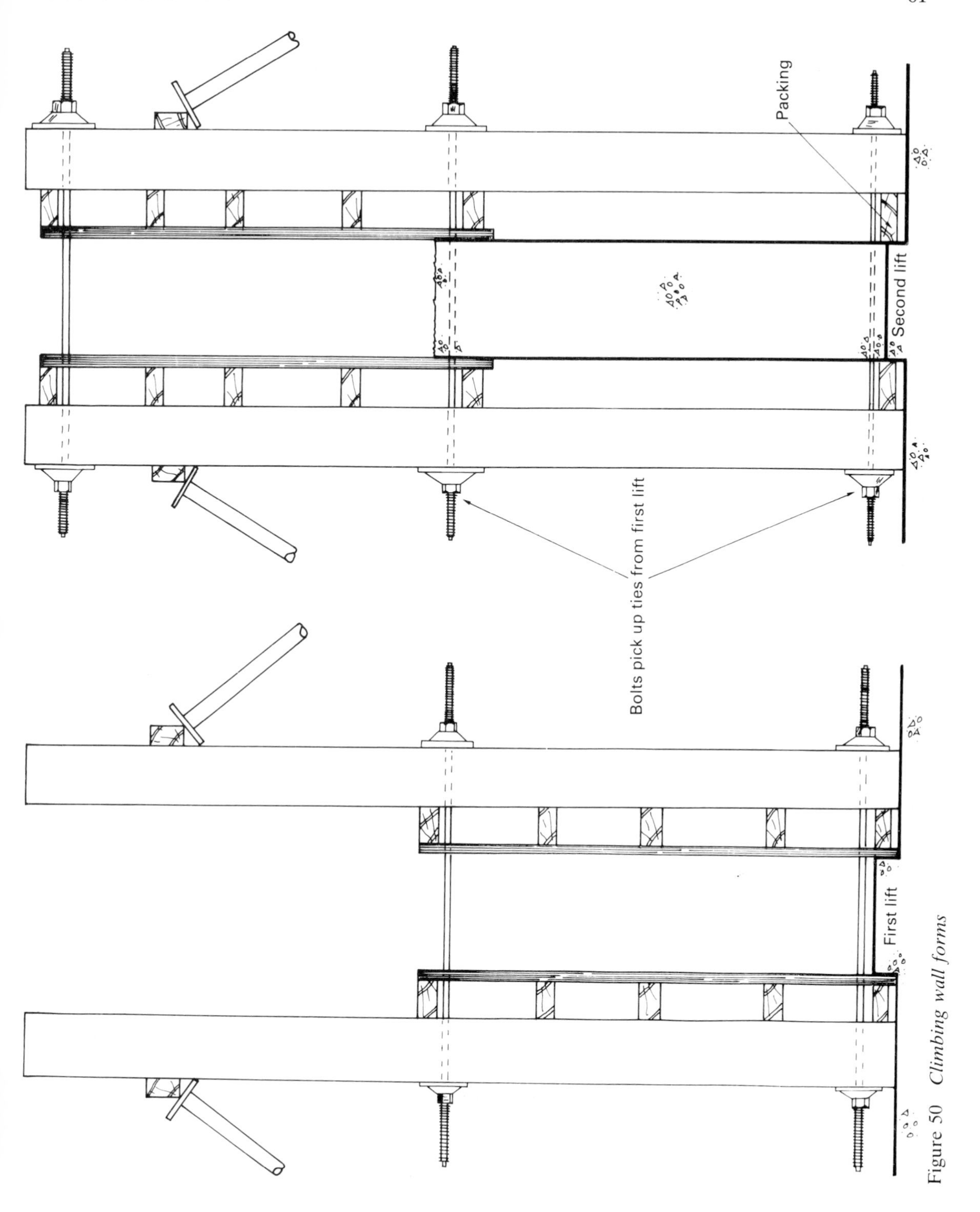

Figure 50 *Climbing wall forms*

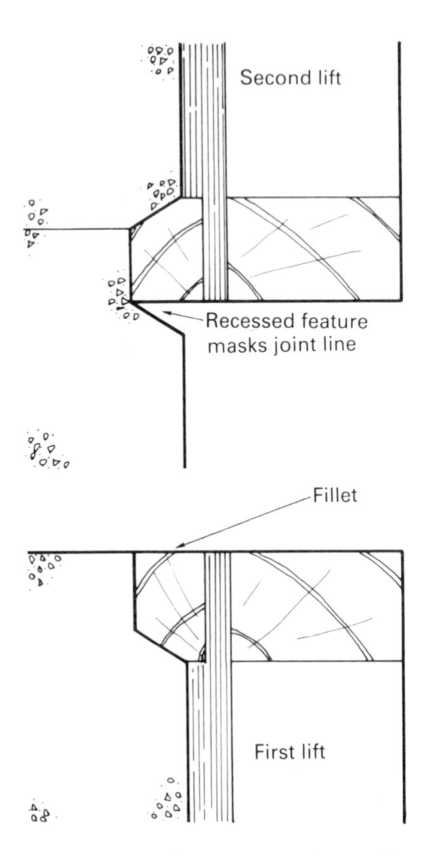

Figure 51 *Use of a feature to mask joint line*

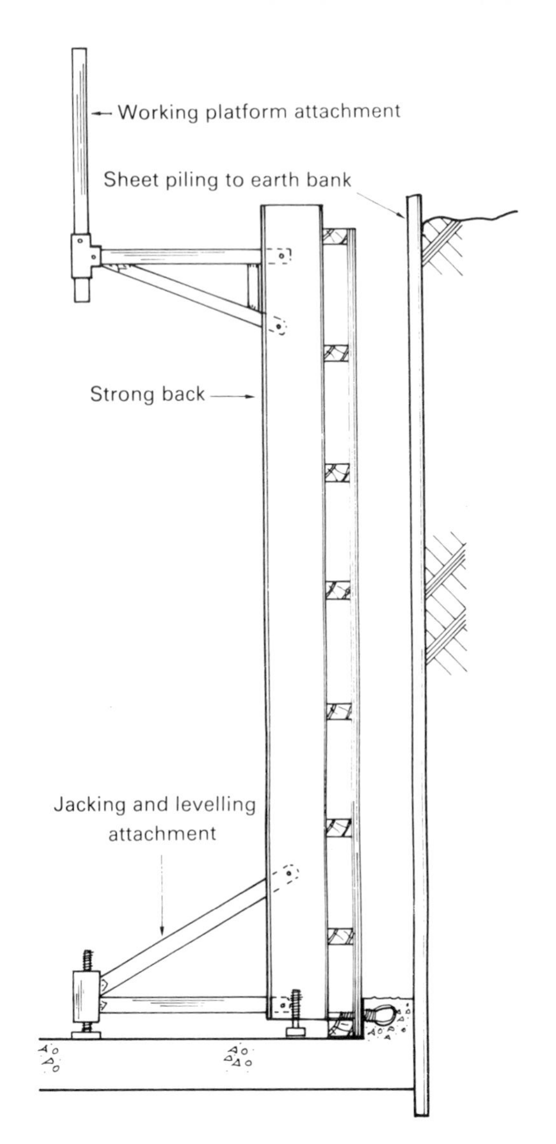

Figure 52 *Single sided wall form*

ness, and therefore will be determined at the design stage when anticipated loadings and pressures are calculated. As a general guide 75 mm × 100 mm joists spaced at 400 mm centres are capable of spanning between ledgers spaced at 1.5 m centres and supported by props at 1.2 m intervals.

Scaffold tube and fittings are used for the lacing and diagonal sway bracing. These permit greater prop loadings and prevent sideways movement of the formwork. The recommended positioning for the lacing members is up one-third of the extended inner tube height, as indicated in Figure 56.

Fork head fittings are located in the top of the adjustable steel props to provide a secure bearing for the ledgers. Most fork heads are of a size to permit the side by side lapping of ledgers.

Concentric (through the centre) loading of props is essential if they are to support their design loading. This can be achieved when fork heads are being used for single ledgers by aligning the square head diagonally across the timber and packing out with wedges (see Figure 57).

Slab edge forms are used where slabs terminate at an open edge. The slab edge form shown in Figure 58 is fixed on top of the decking. It is framed up in timber and plywood and strutted off a ribbon to maintain plumb and line. The strut

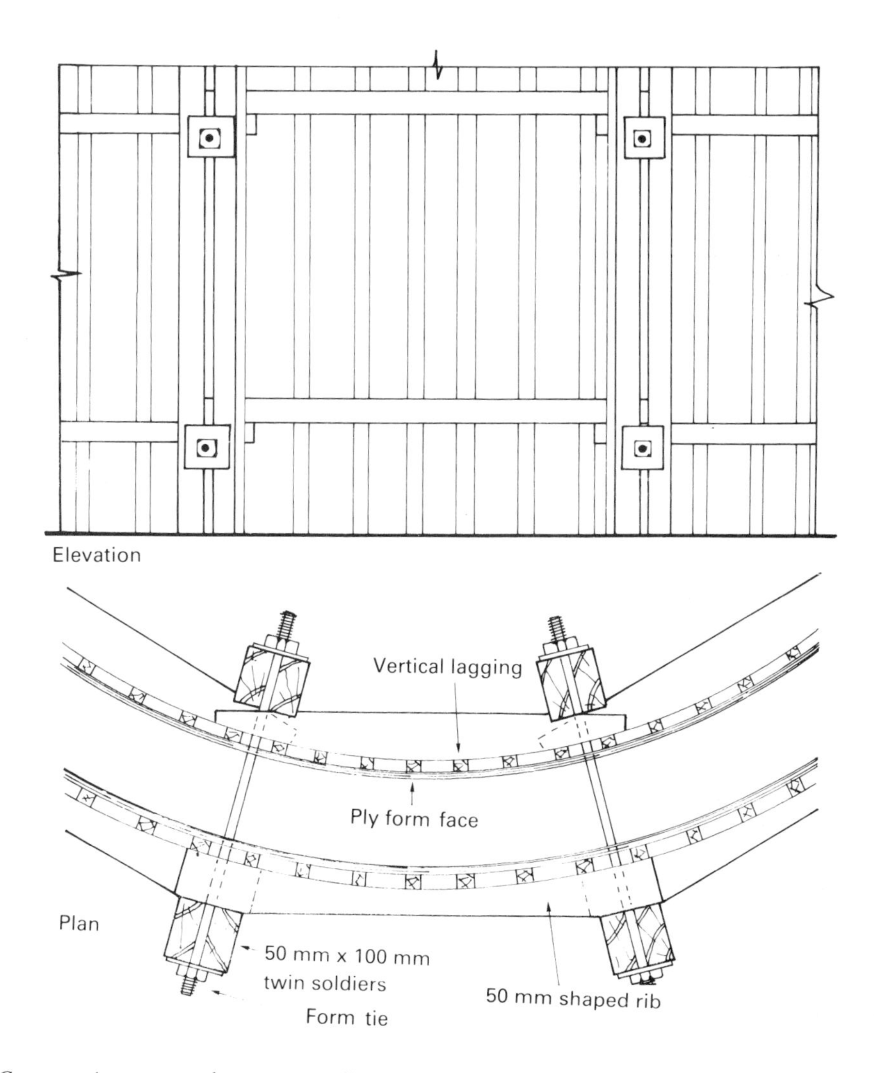

Figure 53 *Constructing a curved concrete wall*

ends are shown bird's mouthed over the ribbon and edge framing. Alternatively they may be cut square and hammered down from the top to permit a bite between the timbers, before finally securing with double headed nails.

Drop beams are sometimes incorporated at slab edges as shown in Figure 59. The beam soffit has been made up using the same method as for slab formwork. The ends of the main slab joists are supported by packing off the beam soffit. The joists supporting the beam soffit have been cantilevered out to enable strutting of the beam side. The cantilevered joists should in turn be strutted back to the adjustable steel props for support. This cantilevered arrangement is often necessary at upper floor levels where there is no direct bearing available for the adjustable steel props directly below the support position.

Figure 60 illustrates a formwork detail where a beam is incorporated in the middle of a floor slab. Where the beam is deep the beam sides may be held in position by the use of form ties spaced

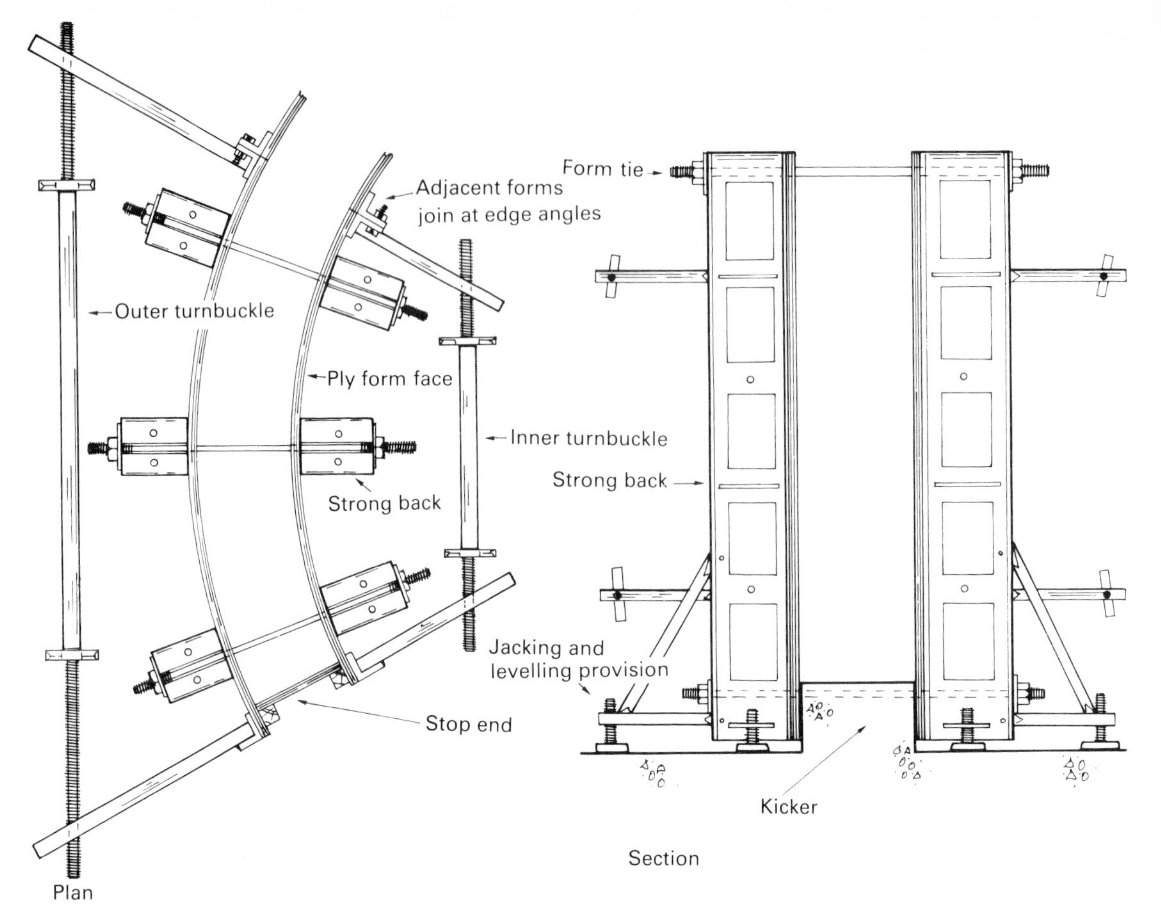

Figure 54 *Proprietary system for forming curved wall shapes*

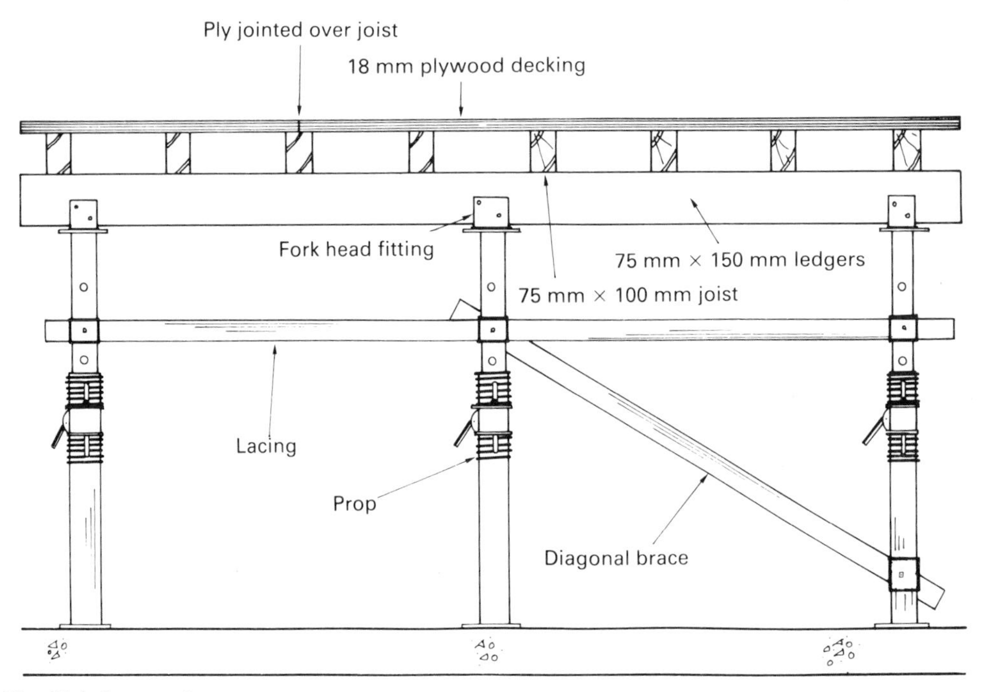

Figure 55 *Slab formwork*

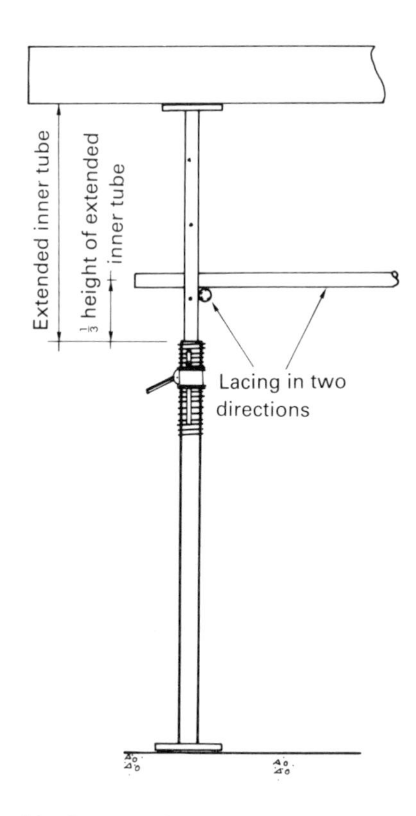

Figure 56 *Lacing of props*

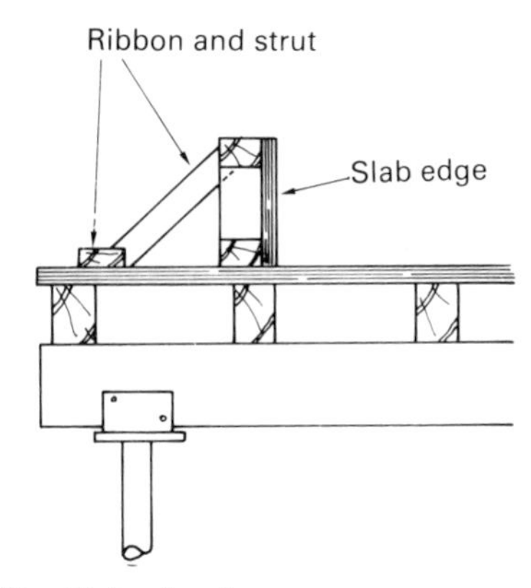

Figure 58 *Slab edge form*

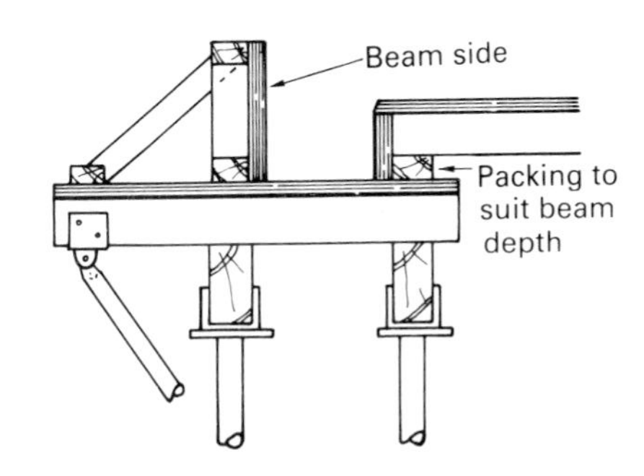

Figure 59 *Drop edge beam*

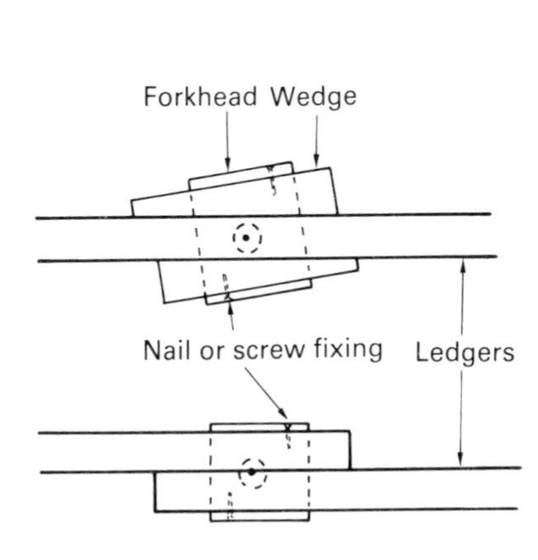

Figure 57 *Concentric loading*

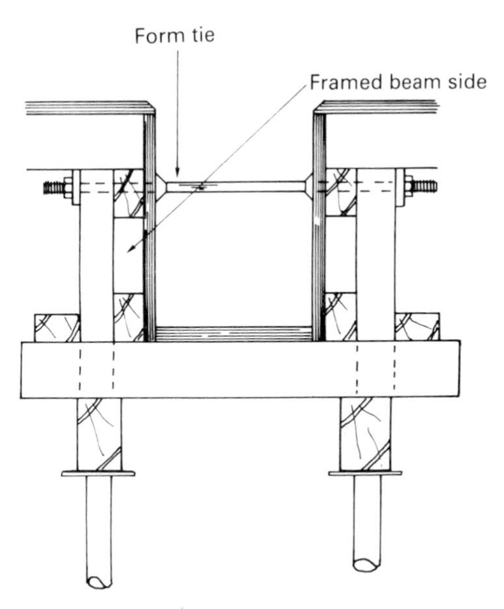

Figure 60 *Drop beam in slab*

periodically along the beam and positioned towards their top edge.

Careful consideration must be given when determining the layout of plywood soffits where they abut previously cast work or existing structure. This is to avoid any trapping after casting. Figure 61 shows how the plywood should be cut around a previously cast column to enable striking. For grout tight joints against the column, any gap can be filled with a compressible foam or rubber strip. Figure 62 shows how narrow striking strips may be included along the edges of slabs where they abut beams or walls. On striking, the splayed edge main plywood sheet comes away first, leaving the narrow striking strip adhering to the soffit. This itself is then simply removed by the insertion of a wedge if required. The ends of the joists are also splayed to prevent binding when one end is lowered.

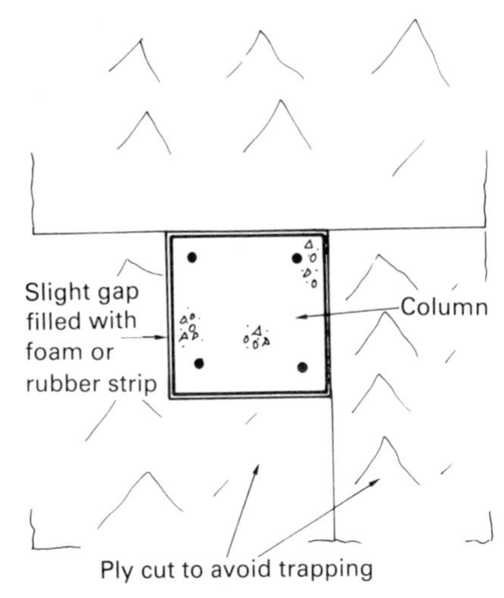

Figure 61 *Cutting plywood around columns*

Beams and lintels are sometimes required to be cast independently of slabs. Figure 63 illustrates a typical formwork detail suitable for the independent casting of a simple beam or lintel spanning between two supports. This detail is very similar to beams cast integrally with slabs. The runners under the beam soffit are not an essential requirement, although their use does enable wider joist spacings and permits easy beam depth adjustment by variation in runner size. The framed beam sides may be strutted as shown, or may be secured using either form ties or beam clamps.

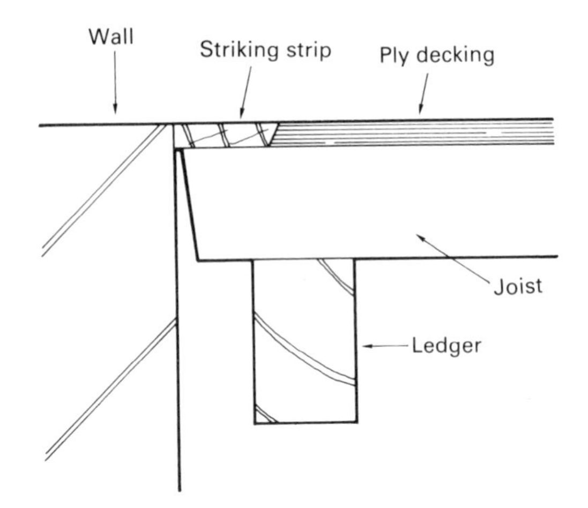

Figure 62 *Use of a striking strip*

Where steel framed buildings are encased in concrete as a corrosion and fire resistance measure, it is often convenient to suspend the majority of the formwork from the steel structure itself. This has the economic advantage of drastically reducing the amount of formwork support structure required, and in addition creates fairly unobstructed under-slab areas. Figure 64 shows a slab and beam formwork arrangement that is suspended off the steel framework using U shaped hangers and she bolts. Concrete spacers are positioned between the steel beam and its formwork soffit. This enables the formwork to be tightened in position whilst at the same time maintaining the required steel cover.

Proprietary adjustable floor centres and standard form panels are useful for decking. As these replace the majority of ledgers, they not only reduce the amount of timber and supports required and so create a fairly unobstructed area, but also greatly speed up the erection and striking operations. Figure 65 shows a typical example of slab and beam formwork utilizing adjustable centres and standard form panels. Adjustable hanger brackets may be used to support edge ledgers in place of the adjustable props where slabs abut previously constructed brick walls. On

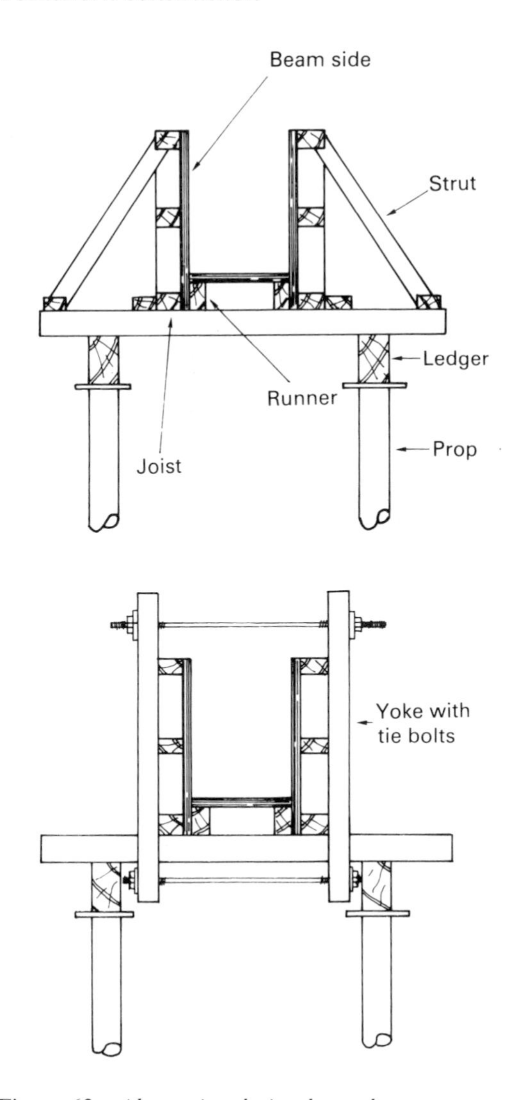

Figure 63 *Alternative design beam boxes*

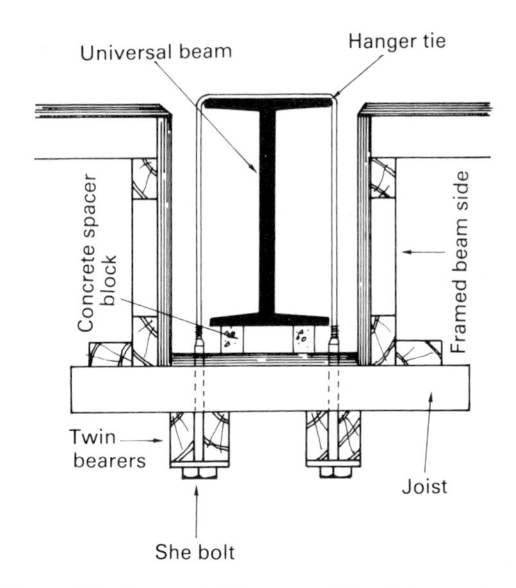

Figure 64 *Suspended beam and slap formwork*

striking, the centres are retracted to free one end and then simply lowered. Temporary props should be positioned under each form panel to prevent them crashing down. Each prop may be removed in turn and the form panel it supports safely lowered (see Figure 66).

In situations where early striking of formwork is required, drop heads may be incorporated either at the top of adjustable props or as part of the proprietary support system. Their use enables the decking to be removed three to four days after curing the slab; the support structure remains undisturbed until the concrete has gained sufficient strength to be self-supporting over its full span. Figure 67 illustrates drop heads fixed to the top of adjustable props. These are used to support proprietary steel beams and standard form panels. On striking the drop heads are released using a hammer blow to move the supporting flange to one side. The panels and beams will drop 100 mm or so, and from this position they can easily be manhandled and lowered.

For repetitive casting of floor slabs in medium to high rise structures, the use of proprietary table form systems can be an economic consideration. These enable erection, striking and subsequent re-erection operations to be carried out with the minimum of dismantling, thus greatly speeding up operations. Figure 68 illustrates a typical table form for slab casting. On striking, the supporting legs are partially retracted to release it from the concrete surface. From there it can be moved sideways, partially overhanging the building, and picked up by the crane for repositioning at the next lift. Where edge beams are incorporated it is essential that the table is capable of being lowered sufficiently to clear the beam soffit.

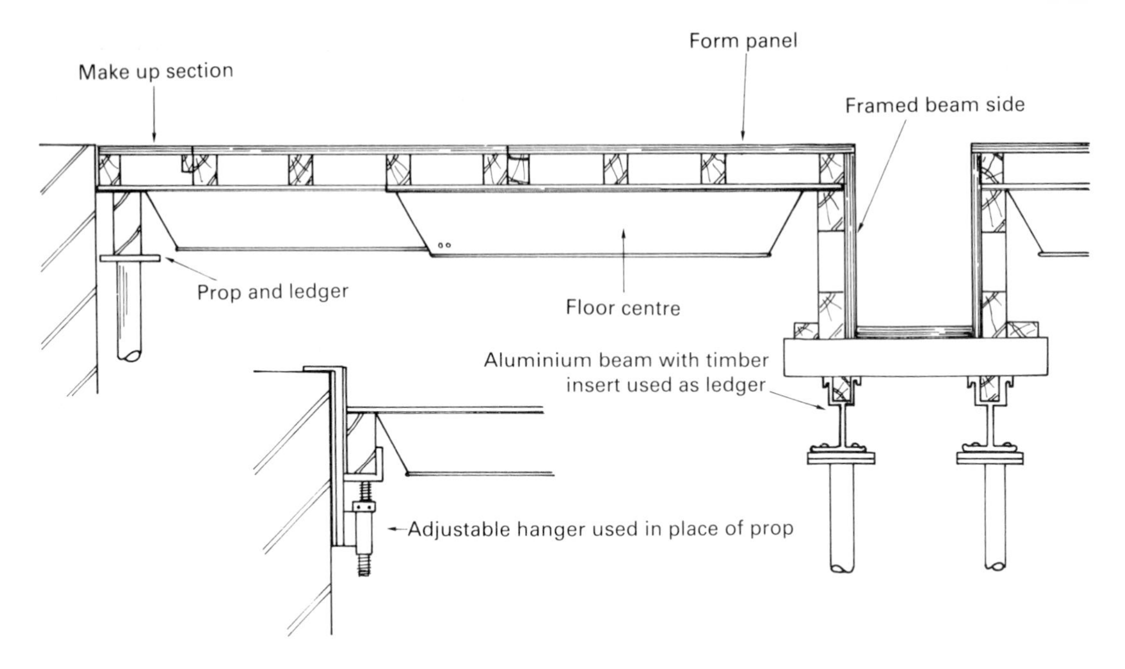

Figure 65 *Use of floor centres*

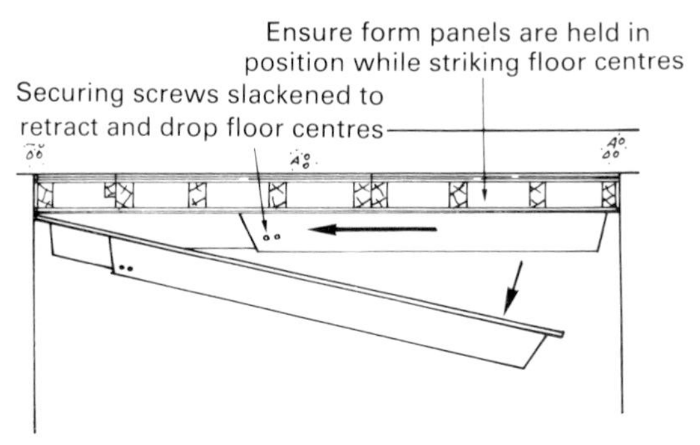

Figure 66 *Striking floor centres*

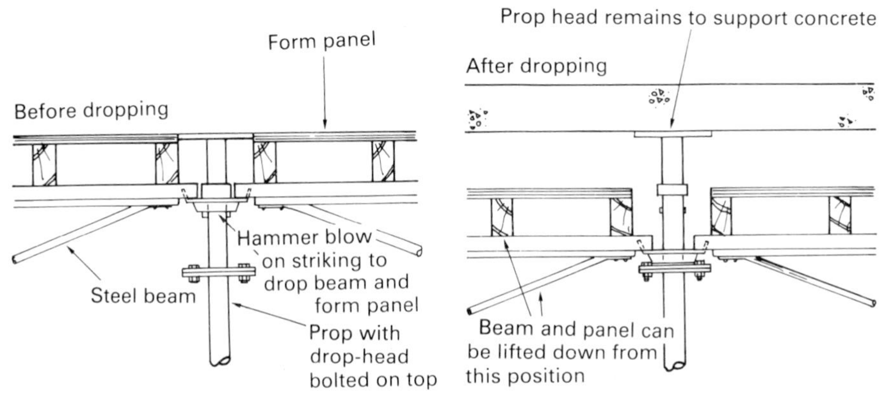

Figure 67 *Use of drop heads*

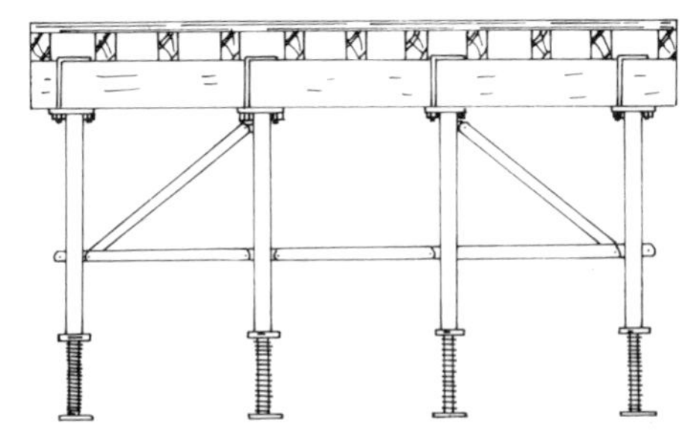

Figure 68 *Slab table form*

Where large open slab spans are required the self-weight of the structure may be reduced by the use of ribbed slabs. These ribs may be formed using proprietary waffles or trough moulds. In addition to reducing the slab mass, a pleasing soffit appearance is created. Figure 69 shows how waffles and troughs may be supported using either proprietary systems or traditional timber ledgers and adjustable steel props. On striking the support structure the moulds can simply be levered from the surface and lowered down. Some waffle and trough moulds incorporate an air valve to enable the connection of a compressed air hose which eases their removal.

Slabs with sloping soffits for use as projecting canopies etc. may be formed using framed-up ribs acting as ledgers to support the joists and plywood decking, as shown in Figure 70. Cross-braces are fixed to the struts of adjacent ribs to prevent overturning.

Alternatively sloping soffits may be formed as illustrated in Figure 71. Here the straight ledgers are raked using differing prop lengths. Wedges cut to the required slope are positioned in the fork head to permit a flat bearing.

A similar method to that of sloping soffits can be used when forming curved slab soffits. Figure 72 illustrates the use of both curved rib supports similar to the ribs of arched centres used for brickwork, and also curved top ledgers for low rises.

No top form is required unless the greatest angle of the curve exceeds about 30 degrees. In this case the top form may be made and fixed in the same way as curved wall forms, using shaped

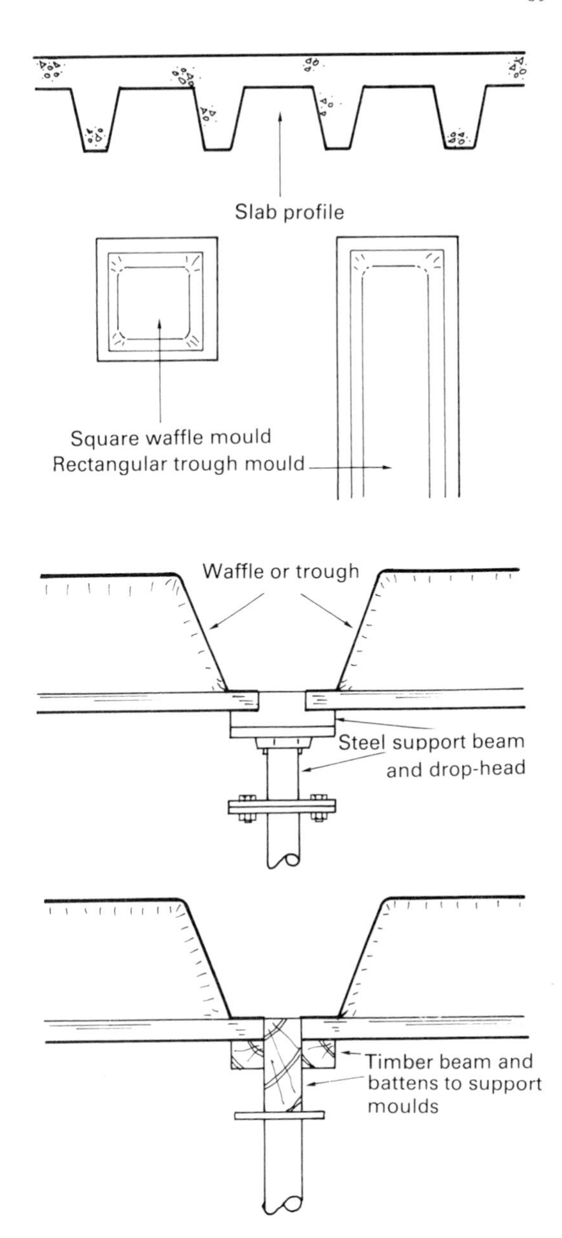

Figure 69 *Waffle and trough details*

ribs and laggings, and secured back to the slab soffit using form ties.

Arched soffit beams forming concrete arches can be cast using a centre made from shaped ribs to form the arched soffit, and standard form panels and soldiers secured with form ties as the

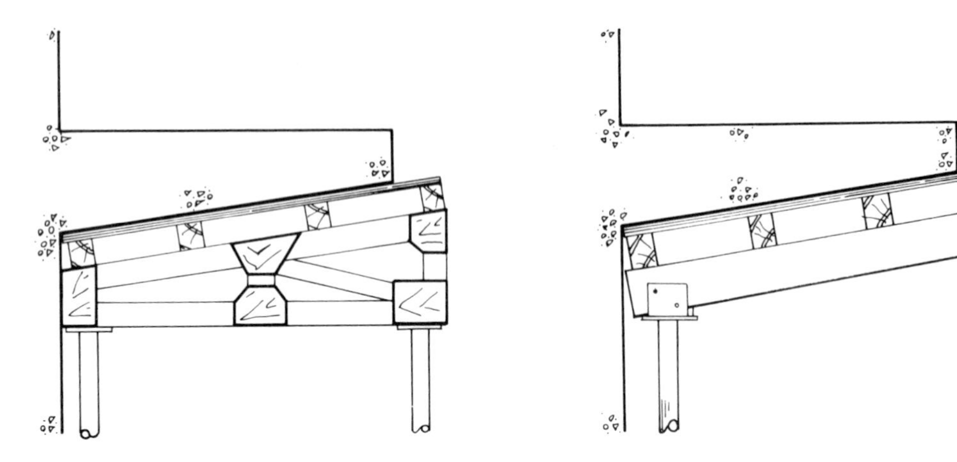

Figure 70 *Sloping slab soffit*

Figure 71 *Alternative sloping soffit detail*

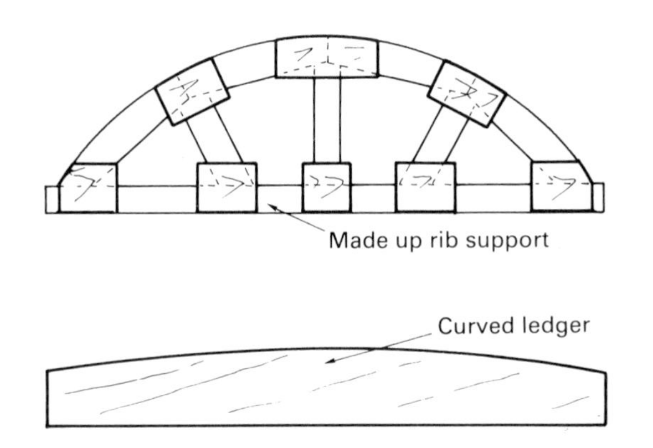

Figure 72 *Ledgers for curved slab soffits*

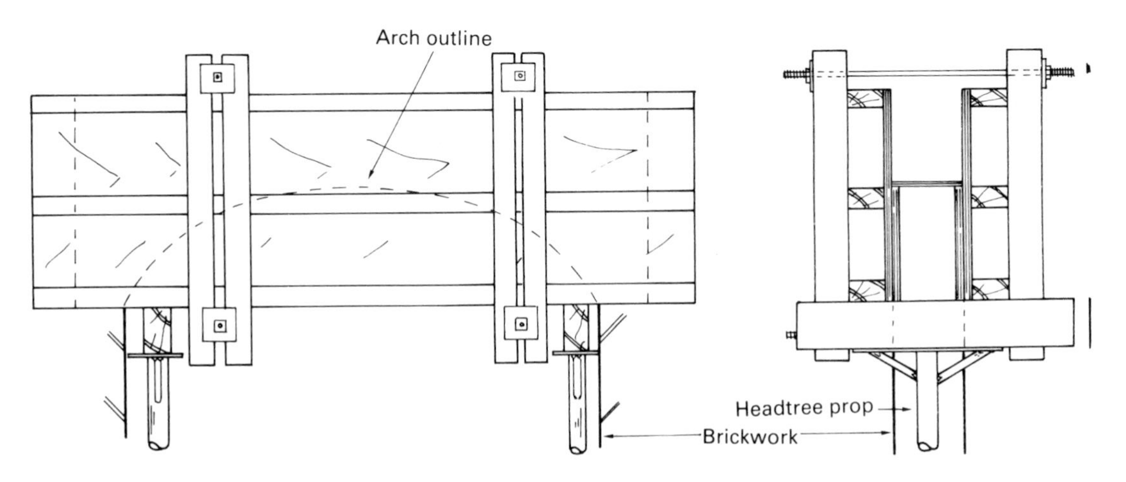

Figure 73 *Arch soffit beam*

Table 18 *Stairs*

Term	*Description of stairs and requirements* *Private stair* *A stairway used by one dwelling only*	*Common stair* *A stairway used by two or more dwellings*	*Stairway in institutional building*	*Stairway in assembly area*	*All other stairways*
Pitch 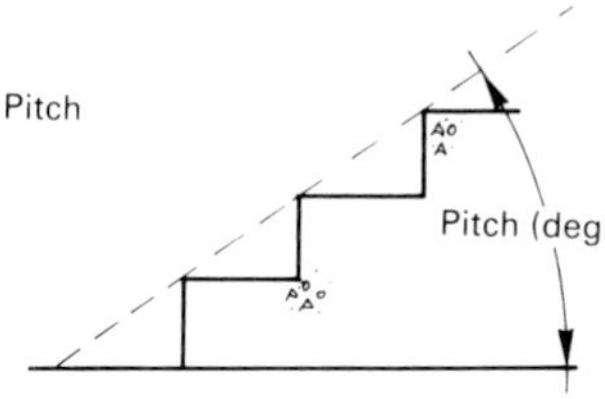 (Pitch (degrees))	42° maximum	38° maximum			
Rise and going 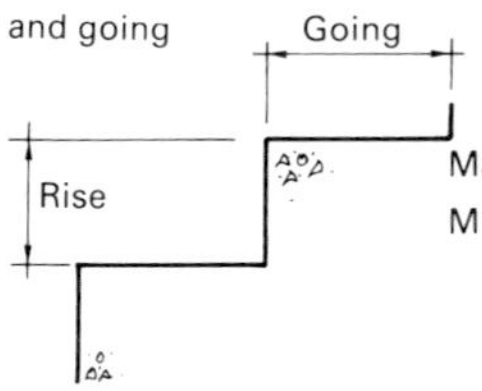 (Going, Rise) Maximum rise	220 mm	190 mm	180 mm	180 mm	190 mm
Minimum going	220 mm	240 mm	280 mm	280 mm	250 mm
	Twice the rise plus once the going ($2R + G$) must in all cases fall between 550 mm–700 mm				
Headroom (Headroom, Pitch line, Headroom) 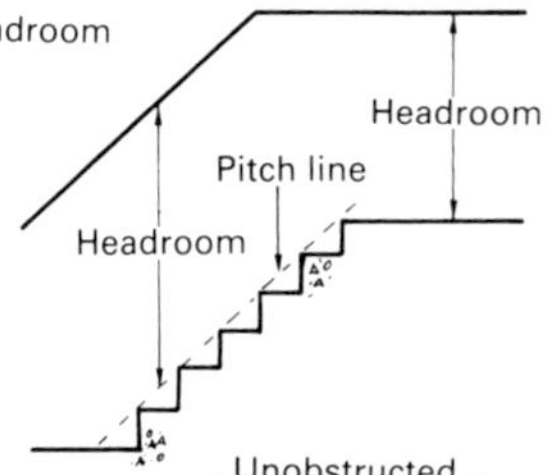	Minimum of 2 m in all cases measures vertically on both flight and landing				
Unobstructed width 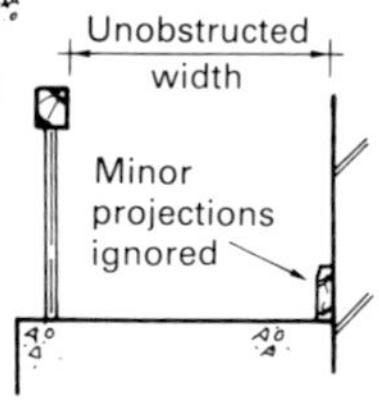 (Unobstructed width, Minor projections ignored)	800 mm	900 mm	1000 mm	1000 mm	800 mm
Landings 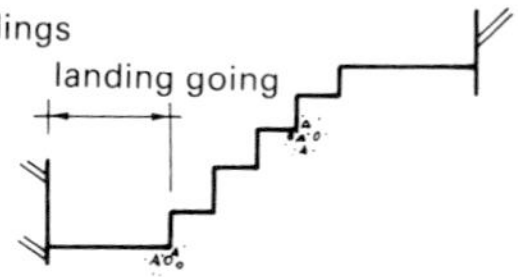 (landing going)	Level landings having a going at least equal to the width of the stair must be provided at each end of a flight				
Length of flight	No more than 36 risers are permitted in consecutive flights unless there is a change of direction of at least 30°. Flights in a shop or assembly area are limited in length to a maximum of 16 risers				

side forms. Details of this arrangement are shown in Figure 73.

Stairs

The provision of stairs in all types of buildings must comply with the requirements of the Building Regulations. These lay down restrictions on pitch, tread and riser size, stair width, headroom etc. for various categories of stairs. Table 18 defines these terms and summarizes the Building Regulation requirements applicable to them. Obviously stair formwork must be designed with this information in mind so that the finished result complies.

Stair soffits can be considered as a sloping slab; thus the supporting formwork will be similar to that of floor slabs. The step profile is formed using cut strings and riser boards. Risers may be formed from timber, plywood or pressed steel as shown in Figure 74. Timber or plywood risers should have a splayed bottom edge to permit the entire tread surface to be trowelled off. The radius along the edges of the pressed steel risers acts as a stiffener and also forms a neat pencil round to the finished step profile.

Figure 75 illustrates two alternative formwork arrangements which are suitable for casting stairs abutting a wall on one side. For stairs that are built between two walls the strutted open edge form can be omitted and substituted by either a wall plate and hangers or a cut wall string. Likewise, free-standing stairs open on both sides can be formed using strutted edge forms on both sides.

Figure 76 shows how cut strings may be marked out using a tread/riser template that has been cut to the desired step profile. This is stepped along the string edge the required number of times and marked around with a pencil each time. The outline of steps is best marked out on an abutting wall from the nosing/pitch line. This is an imaginary line that connects the edge of each step. A chalk line is used to mark this on the wall, and the pitch line length (taken from the template) for each step is marked along it. Using a spirit level the horizontal and vertical lines of the step profile can be marked out from these positions. The thickness of the slab concrete is termed the *waste* (see Figure 77).

Sectional details showing formwork arrangements for short flights either up to or from landings are illustrated in Figure 78. An inverted carriage piece strutted off the main structure should be fixed down the centre of wide flights to resist the tendency of long slender risers to bulge under concrete pressure (see Figure 79).

Precast concrete moulds

The moulds for precast concrete products may be constructed from steel, glass fibre reinforced plastics, or timber and plywood. The first two are mainly the province of precasting factories, while timber and plywood are the common materials for use on site.

The basis of mould box construction is a level base or bed on which edge forms can be assembled and cast. The jointing of edge forms is best carried out using bevelled housings. These tighten on bolting, ensuring a grout tight joint. In addition this detail is less likely to result in damage at the corner of the concrete when

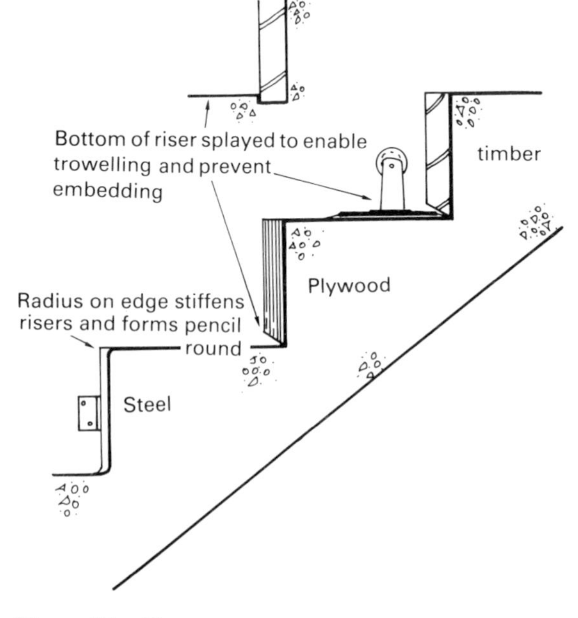

Figure 74 *Risers*

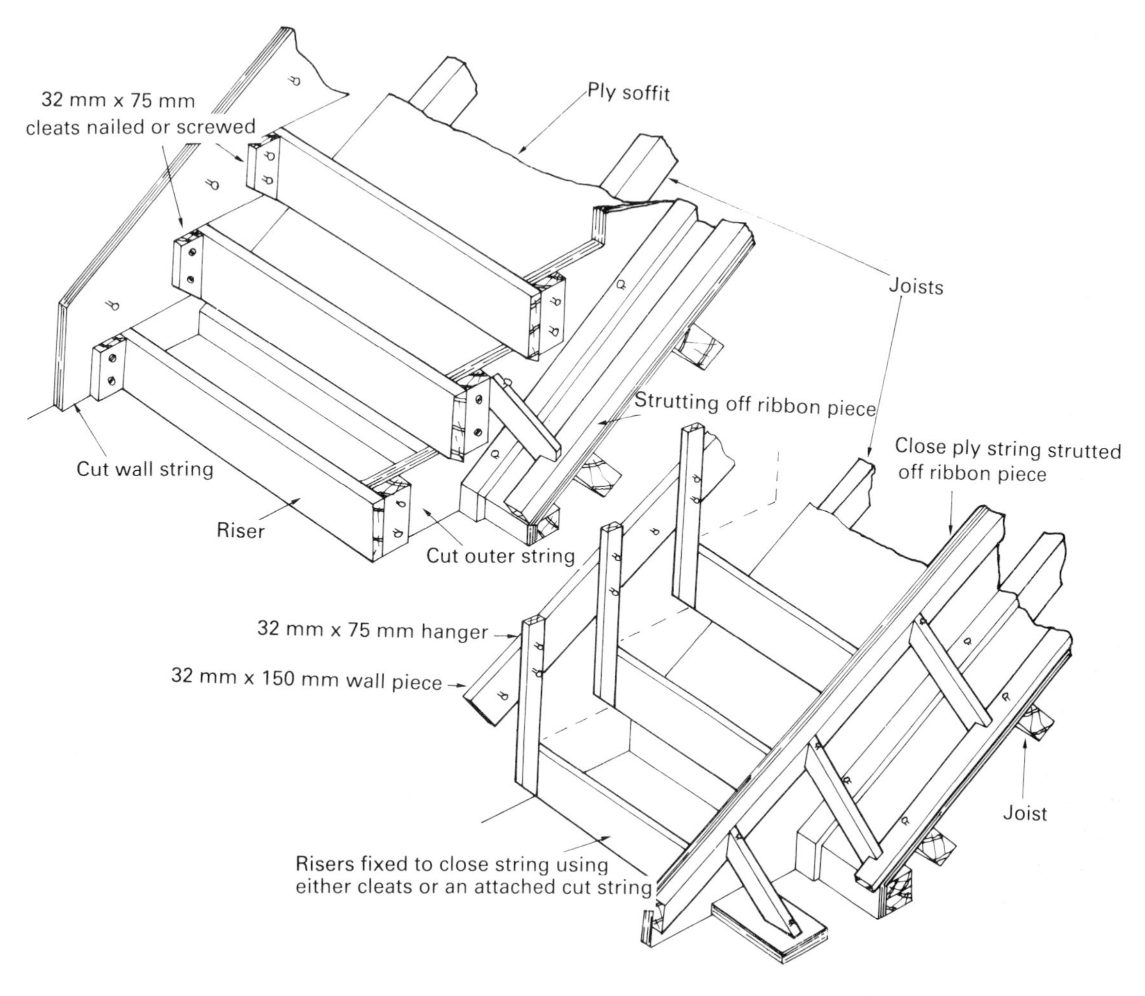

Figure 75 *Casting stairs abutting a wall on one side*

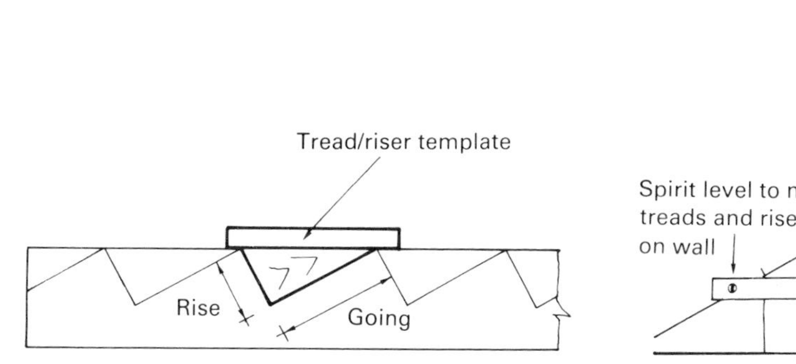

Figure 76 *Marking out a string*

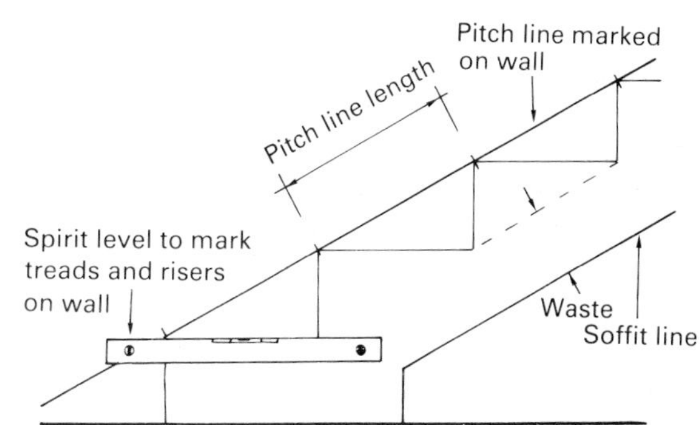

Figure 77 *Marking stair details on abutting wall*

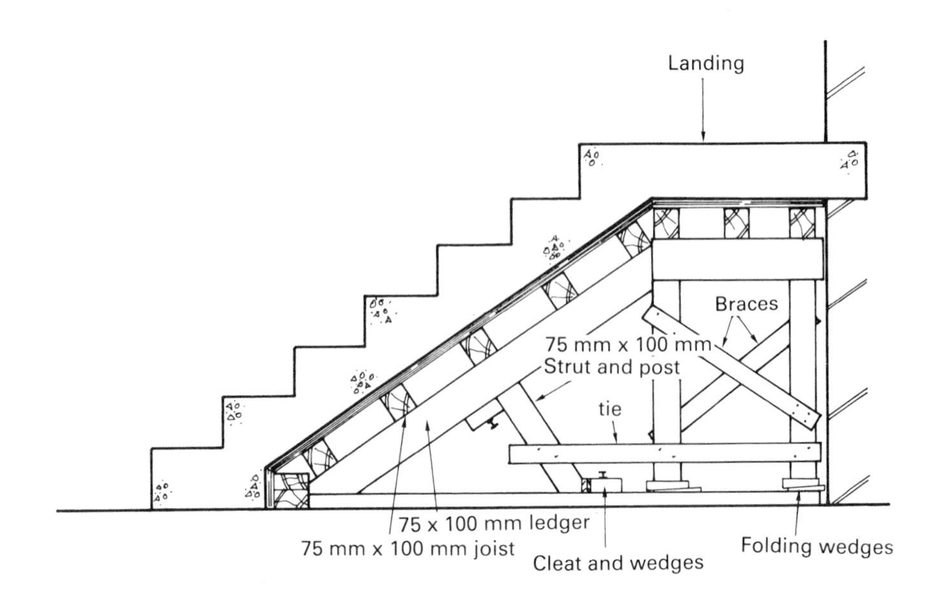

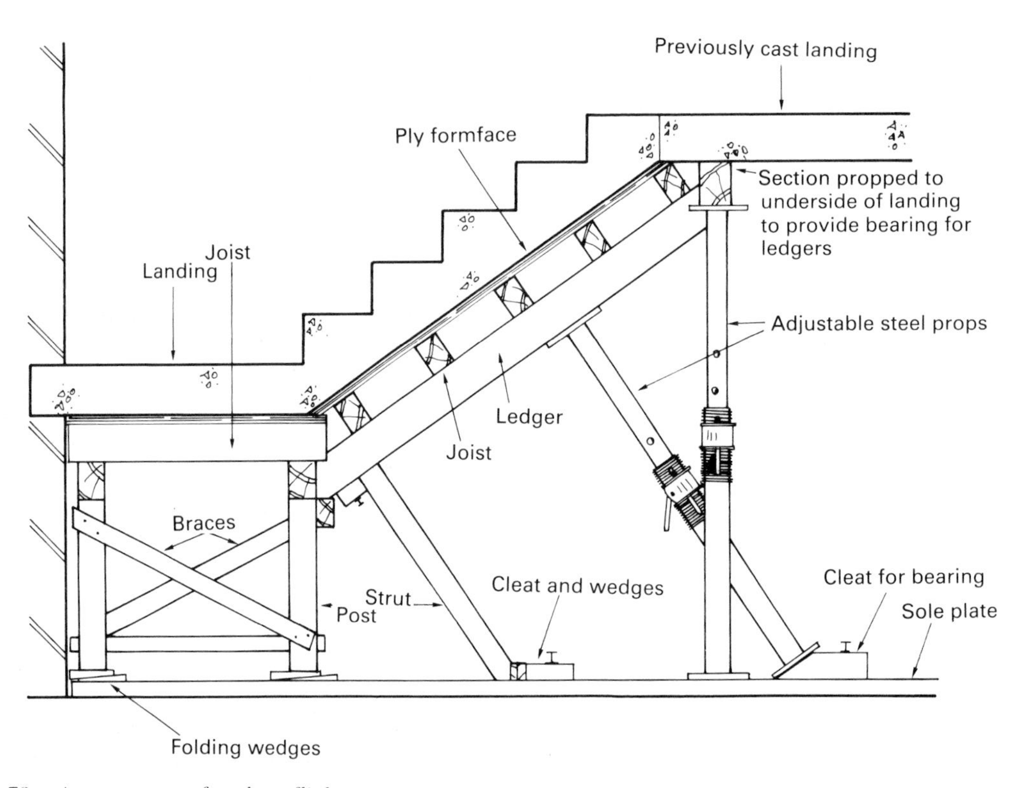

Figure 78 *Arrangements for short flights*

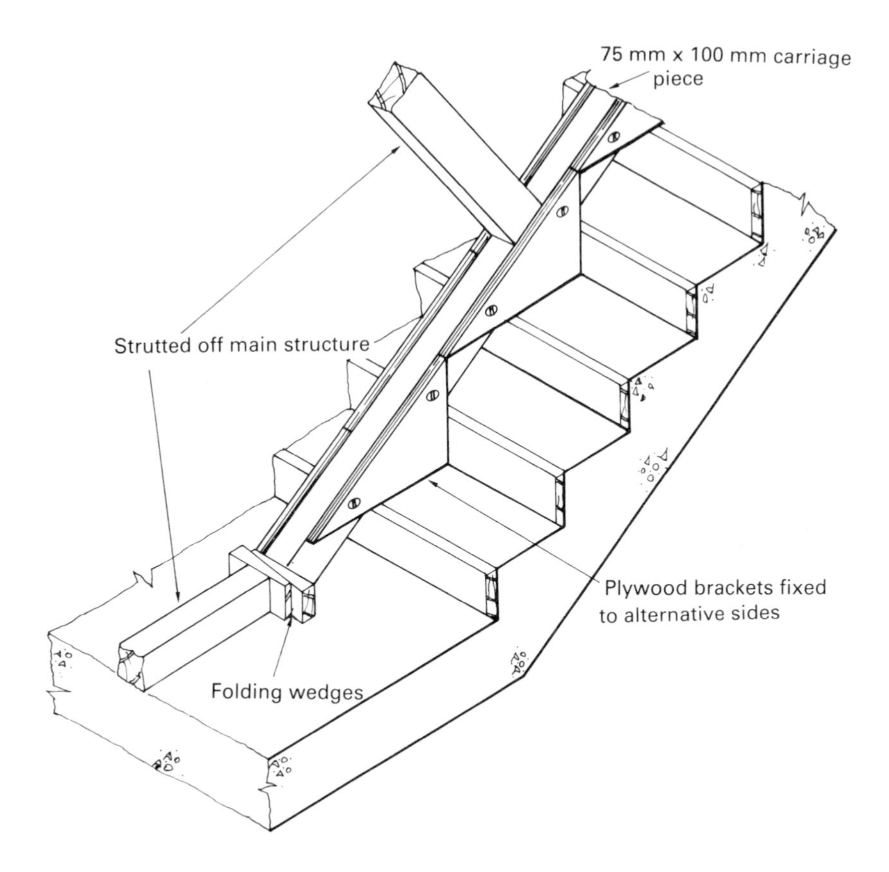

Figure 79 *An inverted carriage piece*

striking. Where not required for wedging purposes, bolt holes for tightening forms are best slotted, since half a turn or so of the nut is all that is required to permit bolt removal. These points are illustrated in Figure 80, which shows a simple mould box suitable for casting blocks, slabs or wall panels.

Moulds that are used to cast more than one concrete unit at a time are termed *gang moulds*. Figure 81 illustrates a typical gang mould for the precasting of tapering fence posts. Holes in the posts for bolts or straining wire are formed by placing well greased steel rods through the distance pieces into the moulds at the required positions. These should be removed shortly after the concrete has commenced stiffening.

Most shaped precast items can be formed by adaptation of the basic mould box design. Figure 82 shows how infill pieces are inserted to a rectangular mould box shape to form a weathered precast sill. For speed of erection and striking, long folding wedges have been used on one side of the box instead of the continuous batten.

Figure 83 shows how a typical variety of concrete shapes can be formed using the infill technique.

In order to facilitate striking, the bed for precast moulds may be pivoted on one edge. After striking the mould box, the bed is tilted to a near vertical position. This causes the cast component to slip a few millimetres on to a restraint batten, and in doing so to break the bond between the concrete and bed face. The concrete

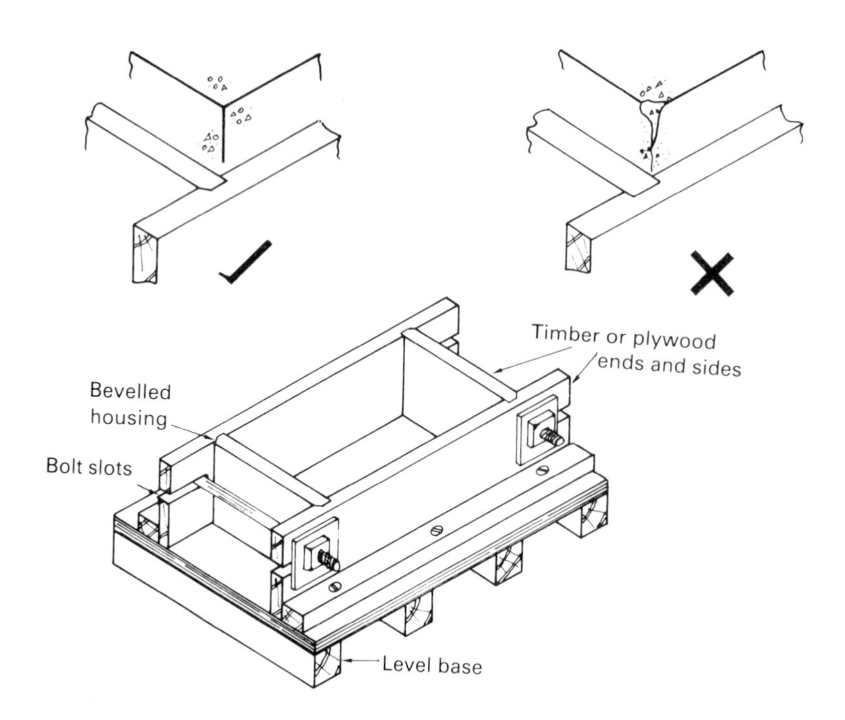

Figure 80 *Basic mould box construction*

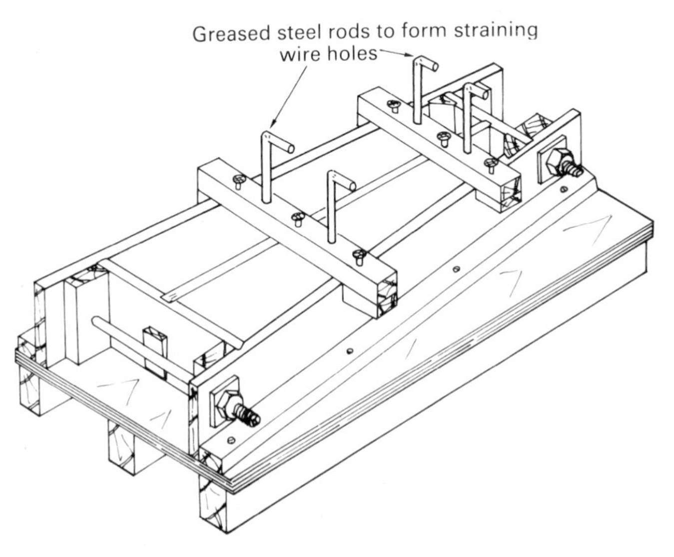

Figure 81 *Gang mould*

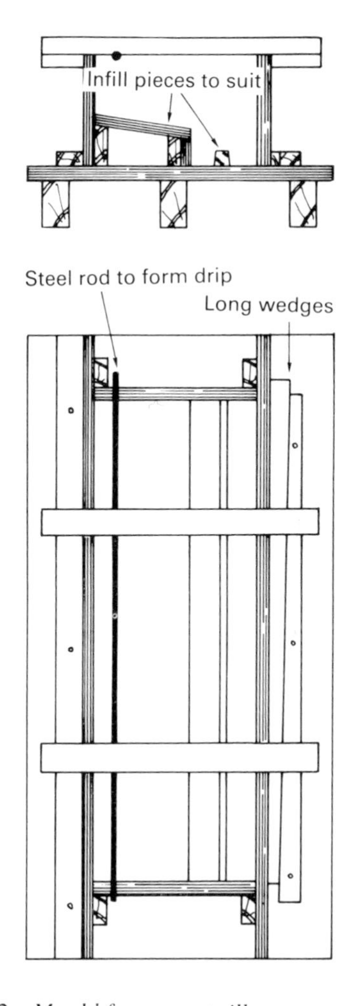

Figure 82 *Mould for precast sill*

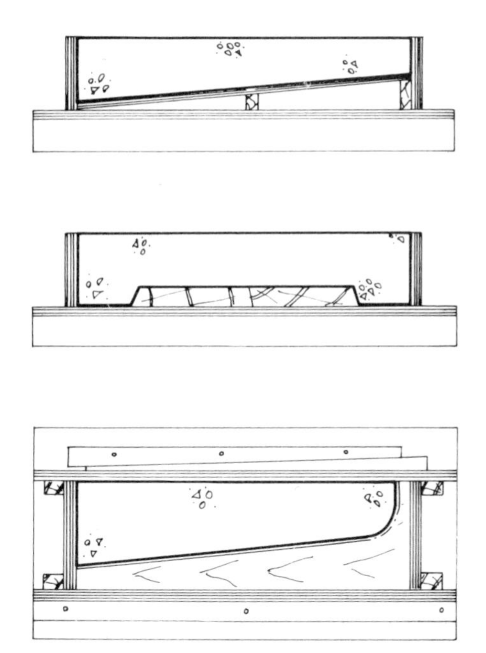

Figure 83 *Infill technique for basic moulds*

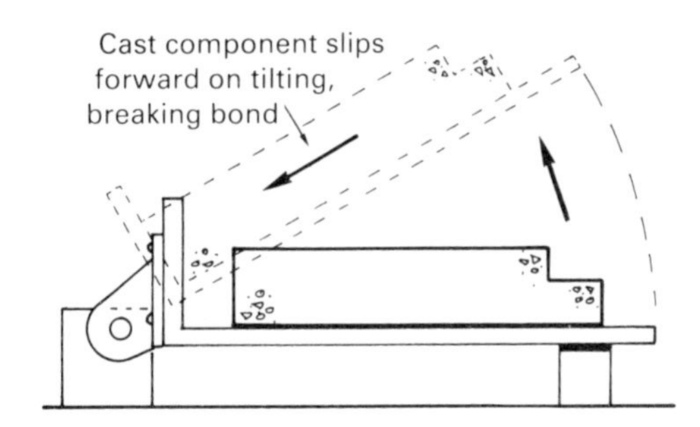

Figure 84 *Tilting precasting bed*

component can them be simply lifted off and transported elsewhere for further curing. This tilting bed method is illustrated in Figure 84.

Stairs are an expensive and time consuming *in situ* casting operation, and so are often specified to be precast. A typical precast stair detail is shown in Figure 85. This incorporates rebates at either end which locate into similar rebates formed at slab landings. Reinforcement loops or cast-in sockets have been provided on the tread surface near the top and bottom of the flight to act as slinging points for crane handling.

Three main casting aspects are possible with stair moulds. These are illustrated in Figure 86. The inclined casting aspect is really an *in situ* casting technique applied to a precast unit. Edge casting is useful for fair faced flights, as all but one edge can have a high quality moulded surface. Inverted flat casting allows the upper surfaces and edges of the flight to be of high class mould quality, whilst the soffit can be trowelled to an acceptable finish. Inverted flat casting is particularly suitable where tiled nosings or non-slip tread surfaces are to be incorporated, as these may simply be placed in the mould prior to filling with concrete.

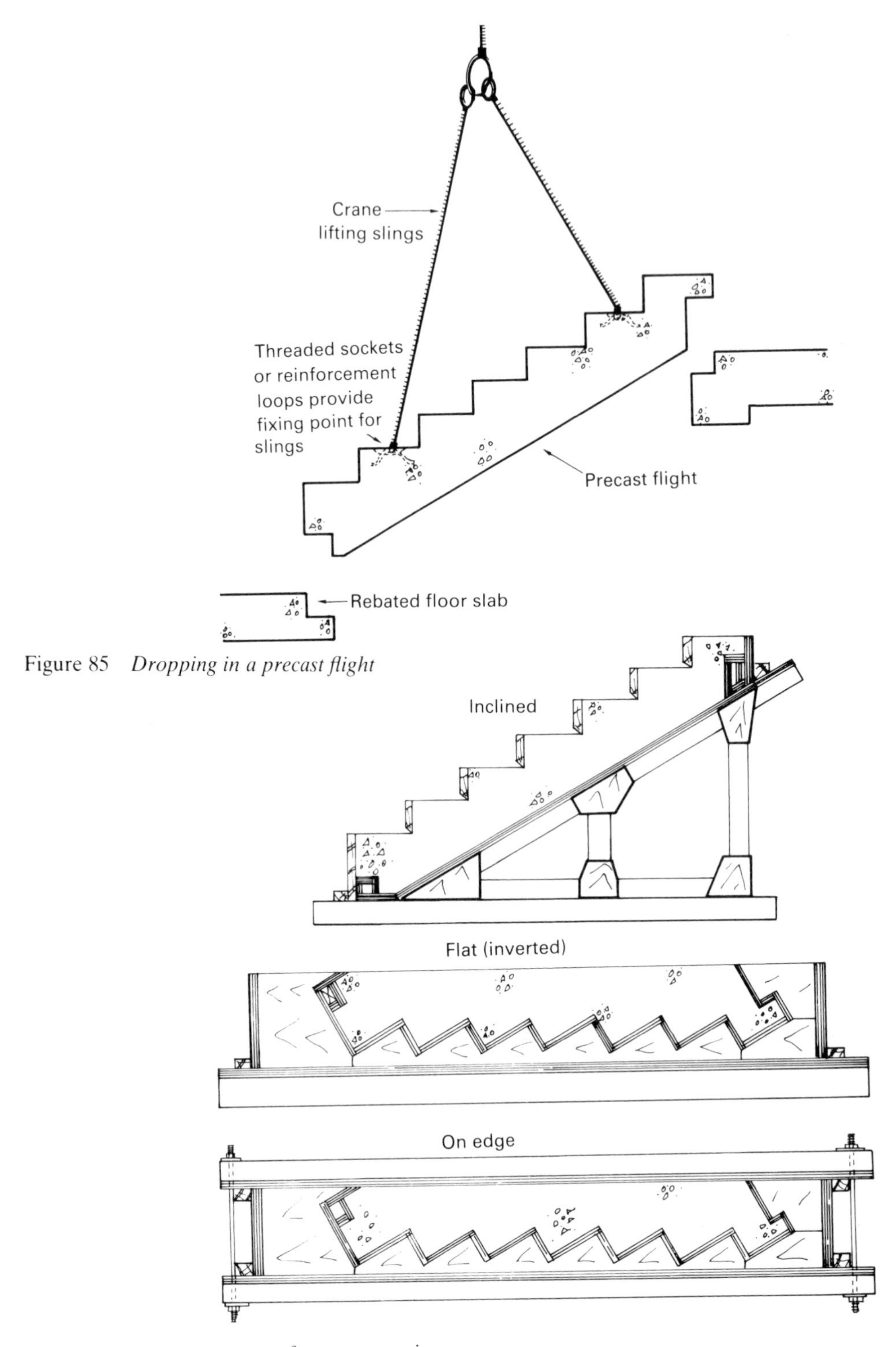

Figure 85 *Dropping in a precast flight*

Figure 86 *Casting aspects for precast stairs*

Self-assessment questions

Question *Your answer*

1 Produce a labelled sketch to show a stop end in a timber and plywood wall form arrangement, including the provision for allowing the steel reinforcement to pass through.

2 Figure 87 is a section through a splayed overhanging canopy. Produce sketches to show a suitable soffit and support structure for its *in situ* casting.

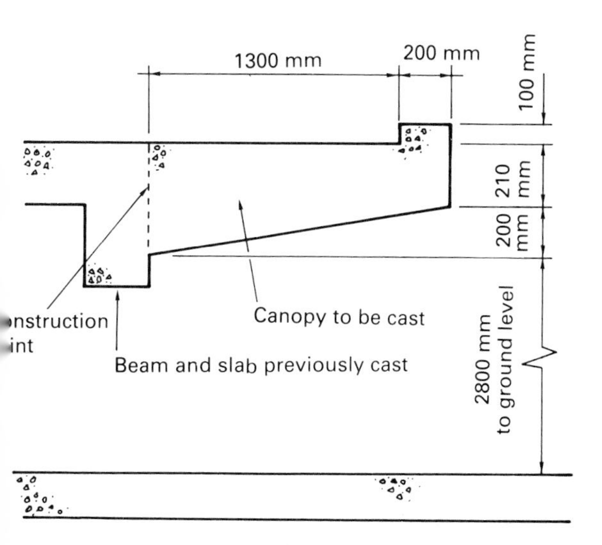

Figure 87 *Splayed overhanging canopy*

3 Figure 88 shows a curved soffit slab which is to form the roof of a pedestrian subway. Sketch a suitable timber and plywood formwork detail.

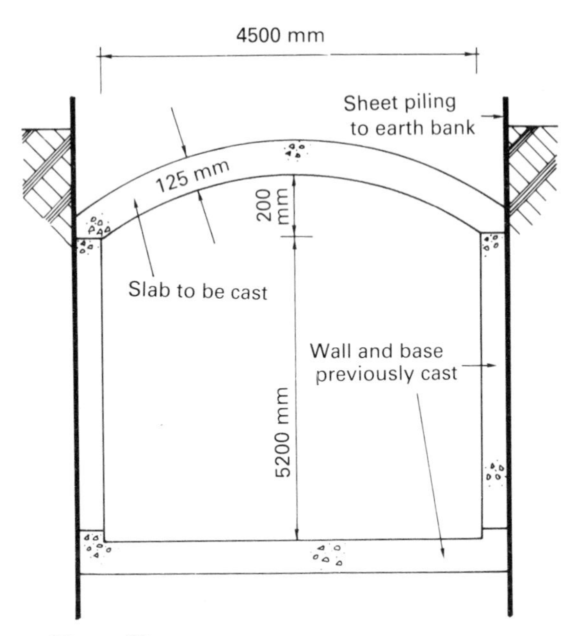

Figure 88

4 Sketch a slab support system that incorporates an early strike facility.

5 Use a labelled sketch to show a box-out suitable for forming a 500 mm diameter opening in a 250 mm thick concrete wall form.

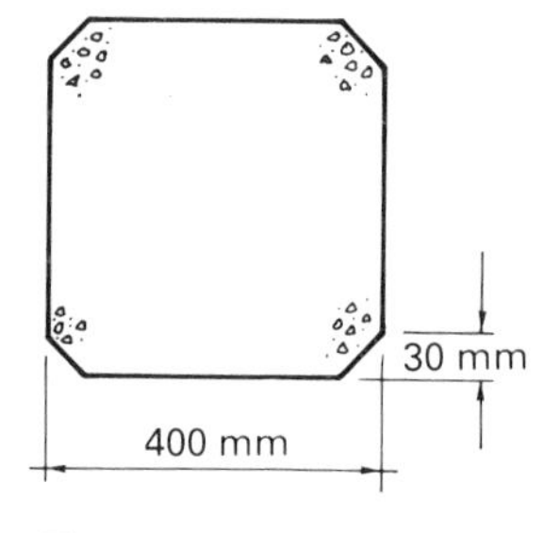

Figure 89 *Column*

6 Sketch a formwork arrangement suitable for casting the horizontal column section shown in Figure 89.

7 Describe with the aid of sketches the use of a kicker in column construction.

8 Sketch and briefly describe the following proprietary formwork items:
(a) An adjustable prop
(b) An adjustable floor centre
(c) A she bolt.

9 Produce sketches to show the construction of a mould box for the on-site precasting of the 1200 mm long concrete window sill illustrated in Figure 90.

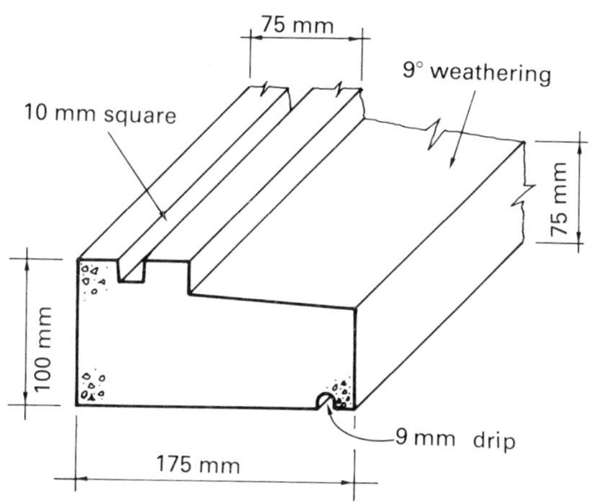

Figure 90 *Window sill*

10 Sketch a cross-section through a 450 mm deep × 300 mm wide beam form supported on adjustable steel props.

Part Two

Concrete

Chapter 4

Properties of concrete

Concrete is a composite material made from a mixture of three basic materials – cement, aggregate and water – and sometimes chemical additives or admixtures:

Cement A fine powder which when mixed with water forms a paste that gradually hardens.
Aggregate A filler material, normally gravel, crushed rock and sand. Classed as:

Fine aggregate (sand, crushed rock fines)
Coarse aggregate (crushed rock, gravel)
All-in aggregate (mixture of fine and coarse aggregates).

Water Normally specified/accepted as being suitable for drinking.

The coarse aggregate forms the bulk of the material; the fine aggregate fills the air spaces (voids) between the coarse aggregate. The cement paste, which is the adhesive, coats the surfaces of the fine and coarse aggregates and bonds them together to form hard concrete.

A typical sample of hardened concrete will consist of 60 to 75 per cent aggregate, with 25 to 40 per cent cement paste and maybe 1 to 2 per cent voids (see Figure 91.) Concrete with a high percentage of cement paste is said to be *rich*, whilst concrete with a low percentage of cement paste is referred to as *lean*.

A chemical reaction called *hydration* commences immediately the cement and water come into contact. Initially the water combines with the surface of the cement grains and the cement paste starts to stiffen. This stiffening process is called *setting*. Gradually the water will penetrate further into the grains and react with the remaining cement, causing the paste to harden and gain strength (Figure 92).

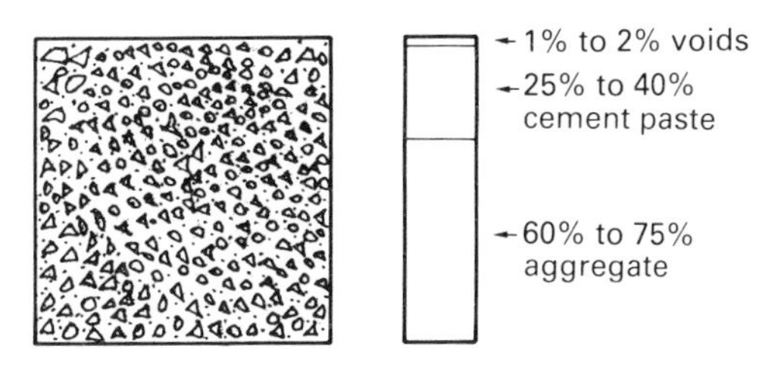

Figure 91 *Typical sample of hardened concrete*

The rate of hydration and thus the gaining of strength proceeds rapidly during the first few days, then continues at a rate which gradually decreases. Normally concrete will obtain about 80 per cent of its ultimate strength in 28 days, although it will continue to gain in strength under moist conditions for 5 or more years. Figure 93 illustrates the strength development of concrete with age. As the time of achieving ultimate strength is uncertain, coupled with the fact that it will be achieved long after the structure is in use, the strength at 28 days is normally used for practical purposes when specifying and assessing the quality of concrete.

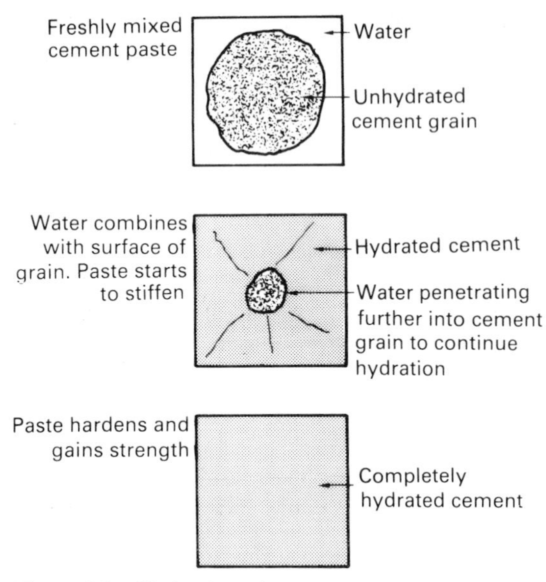

Figure 92 *Hydration of cement*

Types of concrete

Plain concrete A concrete consisting of cement, aggregates and water with or without the addition of admixtures. Used in situations where the nature of loading is mainly compressive, such as mass concrete foundations and slabs.

Reinforced concrete A plain concrete that has had steel embedded into it to increase its tensile strength. Can be used for both compression and tension loading situations, e.g. floor slabs, columns, beams and other structural members.

Prestressed concrete A structural concrete unit or product that has been given a high tensile strength by embedding tensioned wires or cables into the concrete. Pretensioned concrete is formed by casting concrete around previously tensioned wires, whereas in post-tensioned concrete the wires are stretched after the concrete

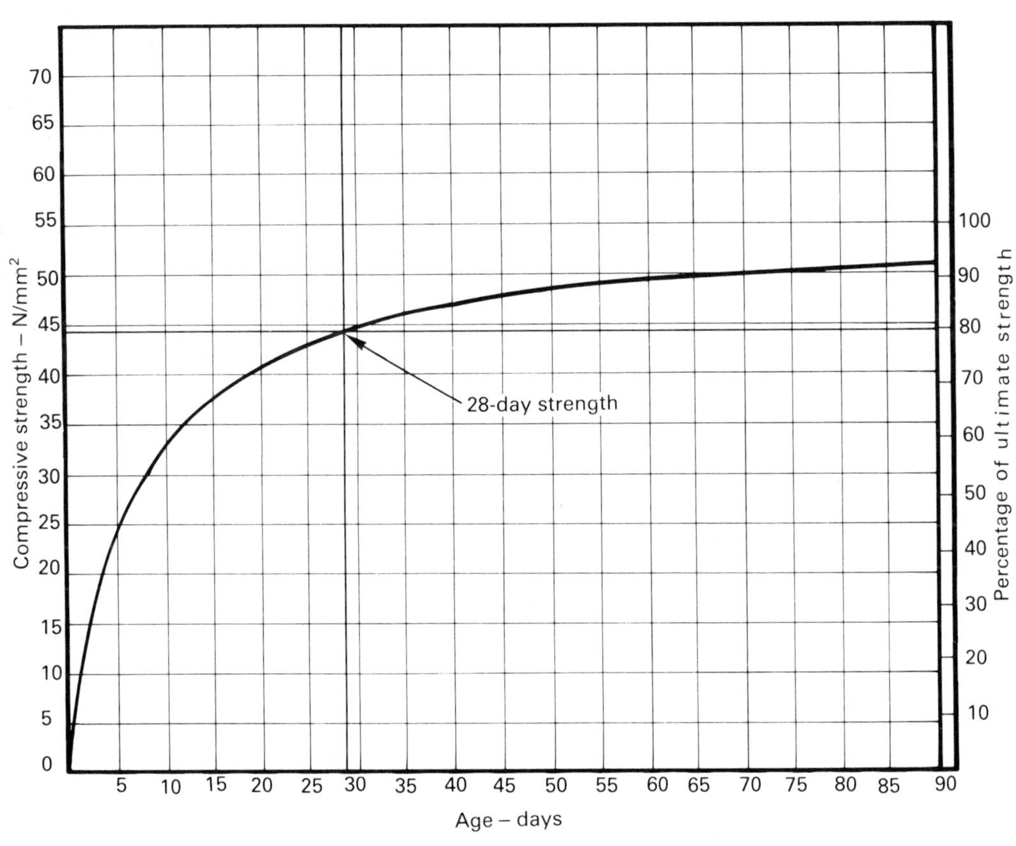

Figure 93 *Typical strength development graph*

has hardened. Pretensioning is used for precast lintels etc., and post-tensioning is used for large span beams and bridges etc.

In situ cast concrete Plain, reinforced or prestressed concrete that has been cast in the actual location where it is required.

Precast concrete Plain, reinforced or prestressed concrete products or units that have been cast away from the location that they will eventually occupy. This could be either in a factory or out of location on the building site.

Lightweight concrete A concrete produced by either aerating the mix or using lightweight aggregates. Used for its light weight, thermal insulation or fire resisting properties.

Properties of concrete

The three main properties of concrete are:

Workability To enable the fluid concrete to be transported, placed and compacted into its required shape.

Strength To enable it to satisfactorily perform its desired structural function.

Durability To ensure that it serves its purpose for as long as possible with the minimum of deterioration.

These three properties are interrelated and are dependent to a large extent upon the quality and correct proportions of the constituent materials in a concrete mix.

The cement in a concrete mix requires a certain amount of water for hydration. Sufficient additional water is required to give it workability; water in excess of this results in air voids. Aggregates should be well graded and the proportions between fine aggregates and coarse aggregates controlled so that, when the concrete is compacted, air voids are kept to a minimum. Sufficient cement paste is required in a mix to coat each aggregate particle and bond the particles together.

Air voids in a mix cause a loss of strength. Interconnected air voids make concrete susceptible to frost damage and may also allow moisture to reach the steel reinforcement, causing it to corrode. Insufficient cement paste in a mix causes a reduction in strength owing to the lack of aggregate bonding.

These factors can be controlled by specifying the following constituent ratios.

Water/cement ratio

This ratio is found by dividing the mass of water in a mix by the mass of cement in that mix:

$$\text{water/cement ratio} = \frac{\text{mass of water}}{\text{mass of cement}}$$

This is also known as the *free* water/cement ratio as it includes the added mixing water, the moisture on the aggregate's surface, but not the absorbed moisture contained in the aggregates pores.

Aggregate/cement ratio

This ratio is found by dividing the mass of aggregate in a mix by the mass of cement in that mix:

$$\text{aggregate/cement ratio} = \frac{\text{mass of aggregate}}{\text{mass of cement}}$$

Fine/coarse aggregate ratio

This ratio is found by dividing the mass of fine aggregate in a mix by the mass of coarse aggregate in that mix:

$$\text{fine/coarse aggregate ratio} = \frac{\text{mass of fine aggregate}}{\text{mass of coarse aggregate}}$$

Example

A typical concrete mix contains the following constituents:

170 kg of water
300 kg of cement
665 kg of fine aggregate
1200 kg of coarse aggregate

The constituent ratios are as follows:

water/cement ratio = 170/300 = 0.567

aggregate/cement ratio = 1865/300 = 6.217

fine/coarse aggregate ratio = 665/1200 = 0.554

Example
Determine the mass of water required to give a water/cement ratio of 0.4 when using 150 kg of cement (three bags).

mass of water = 0.4 × 150 = 60 kg

Example
Determine the mass of aggregate required for 150 kg of cement using an aggregate/cement ratio of 6.

mass of aggregate = 6 × 150 = 900 kg

Example
Determine the mass of fine aggregate for a mix having a coarse aggregate content of 600 kg, using a fine/coarse aggregate ratio of 0.5.

mass of fine aggregate = 0.5 × 600 = 300 kg

Example
Determine the mass of fine aggregate for a mix having a coarse aggregate content of 600 kg, using a fine/coarse aggregate ratio of 0.5.

mass of fine aggregate = 0.05 × 600 = 300 kg

Workability

Workability is the term used to describe the ease with which a concrete mix can be placed and compacted. This is a property of fresh concrete that can range from *high* (a runny free-flowing mix) to *extremely low* (a dry stiff mix).

The three interrelated factors that are considered when determining a suitable workability for a particular situation are:

Method of compaction Higher workabilities are required for hand compacted concrete than when using mechanical vibrators.
Formwork dimensions High workabilities are required for narrow sections and intricate shapes to enable the concrete to flow, whereas stiff low workability mixes may be suitable for large mass concrete sections.
Reinforcement details High workabilities are required where reinforcement is closely spaced to prevent restricting the concrete flow.

The main factors that affect the workability of a concrete mix are as follows.

Constituent ratios
Water/cement ratio A ratio of 0.25 is sufficient for hydration, but unfortunately the mix would be too dry for compaction. Additional water is required to lubricate the aggregate particles and enable the concrete to flow.
Aggregate/cement ratio The lower the ratio the higher the workability, providing the water/cement ratio remains unchanged. This is because in low aggregate/cement ratio mixes there will be additional cement paste for lubrication.
Fine/coarse aggregate ratio The lower the ratio the higher the workability, providing the aggregate/cement ratio remains unchanged. This is because in low fine/coarse aggregate mixes less cement paste is required to coat the larger aggregate particles, and thus more is available for lubrication.

Aggregate size, shape, texture and grading
Large aggregate particles have a smaller surface area than an equivalent volume of smaller particles. Thus the larger ones require less cement paste, leaving more for lubrication and resulting in a higher workability.

Rounded aggregates slip over each other easily, producing high workabilities; angular and cubical aggregates interlock, resulting in a low workability.

Rough texture particles offer more resistance to slip than do smooth texture particles. Thus the workability will be lower when using aggregates with rough texture particles.

Finely graded aggregates offer a greater particle surface area than do medium or coarsely graded aggregates. Thus workability decreases as the grading becomes finer.

Cohesion
This is a property of concrete that is closely related to and interdependent with the workability of a concrete mix, and it is therefore considered under this heading. Cohesion can be defined as the attraction between the constituents

of concrete causing them to stick together. Lack of cohesion results in segregation. High workabilities and rich mixes can lead to wet segregation, while low workabilities and lean mixes can result in dry segregation. Large aggregate particles, being heavier, tend to segregate more than smaller aggregate particles. Rounded aggregate particles have greater cohesion than do angular aggregates. Smooth aggregate particles also have greater cohesion than rough texture particles. Coarsely graded aggregates are more liable to segregate than are medium or finely graded aggregates.

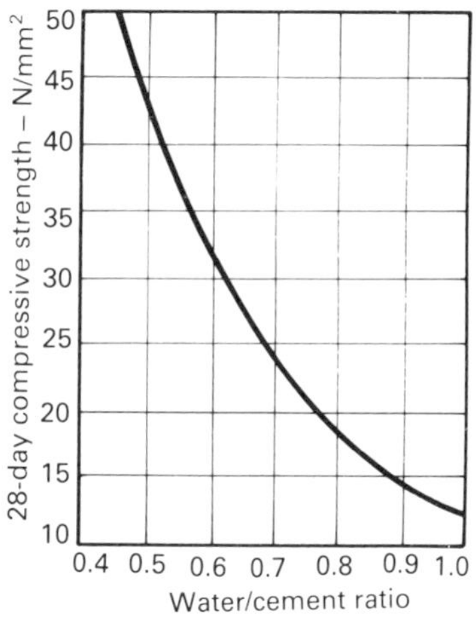

Figure 94 *Variation of strength with water-cement ratio*

Strength

Concrete is normally specified with regard to strength by its compressive strength at an age of 28 days. The main factors that affect the compressive strength at this age are as follows:

Void content The compressive strength of concrete rapidly decreases as its void content increases. A 1 per cent increase in void content can cause a 6 per cent decrease in compressive strength. Thus a concrete sample with a 5 per cent void content could be 30 per cent below strength.

Degree of compaction Full compaction is essential in order to minimize the void content. The method of compaction used is dependent on the degree of workability. High workabilities are suitable for hand compaction, whereas low workabilities require mechanical vibration.

Water/cement ratio A low water/cement ratio may result in a dry low workability mix which is difficult to compact. High water/cement ratios give a concrete mix that has a high workability, making it easy to compact. Excess water, however, will rise to the surface of freshly placed concrete (known as *bleeding*), leaving behind a fine interconnected network of voids. Figure 94 illustrates that strength decreases as the water/cement ratio rises. The strongest concrete is produced using the lowest water/cement ratio that is compatible with the degree of workability required to enable compaction.

Aggregate/cement ratio In high aggregate/cement ratio mixes there may be insufficient cement paste to coat each aggregate particle, resulting in a lack of bond and an increase in voids.

Fine/coarse aggregate ratio and aggregate grading In low fine/coarse aggregate ratio or poorly graded aggregate mixes there may be insufficient fine aggregate particles to fill the voids between the larger particles, resulting in strength reduction.

Aggregate shape and texture Angular rough textured aggregate particles produce stronger concrete than do rounded smooth textured ones. This is due to the interlocking of the angular particles, giving each other support, and the good mechanical bonding between the cement paste and rough textured surface. By contrast, rounded smooth textured particles tend to slip and slide. However, this is partially offset by the increased cohesion between the smooth surfaces.

Curing is the process of retaining water in freshly cast concrete and protecting it from extremes of temperature. The longer this process is continued the more complete will be the hydration, and thus the greater will be the concrete strength.

Durability

In general the durability of concrete increases as its void content decreases. Concrete with a high void content is susceptible to the penetration of fluids. Fluids can freeze and consequently expand, causing a breakdown in the bonding of the surface concrete and eventually leading to spalling; or carry with them weak acid solutions from atmospheric pollution, causing chemical reactions that disintegrate the concrete; or carry with them a sulphate solution from the soil, which again causes a chemical reaction with the concrete, resulting in expansion and disintegration.

In addition, concretes with a high void content are less resistant to abrasion; floor surfaces etc. wear very quickly. They are more susceptible to cracking and spalling due to repeated expansion and contraction caused by alternating climatic conditions, e.g. rain then sunshine. Finally, voids provide a convenient anchorage for the growth of mould and plant life, which is not only unsightly but can also lead to surface disintegration.

Other properties of concrete

Thermal

Air, being a conductor of heat, serves to improve the thermal insulation properties of concrete. Therefore high void content concretes have a better standard of thermal insulation than those with a minimum of voids.

Fire

Concrete is a non-combustible material. However, its resistance to fire is dependent on the aggregate used. Many natural aggregates expand and break up when exposed to high temperatures. Manufactured aggregates that have been produced using high temperatures are more stable. Irrespective of the aggregates used, a reduction in strength will result due to a dehydration of the cement paste at the surface. Apart from this, reinforced concrete will experience a further reduction due to the loss in strength of the steel reinforcement when it is exposed to high temperatures.

Sound

A material's sound or acoustic properties are measured in terms of its ability to resist the passage of sound through it and its ability to absorb sound. High void content concrete and lightweight concretes are good sound absorbers.

Reinforced concrete

Plain concrete is very strong when subjected to compressive stress, but it is comparatively weak in tension. Its tensile strength is about 10 per cent of its compressive strength.

Very few structures are subjected to loadings which are totally compressive in nature. Thus steel reinforcement is introduced to resist the tensile and shear stresses, producing a composite material called reinforced concrete.

Structural loading

Stress

When any structural member is subjected to a force, it is said to be in a state of stress. The types of stress (see Figure 95) are as follows:

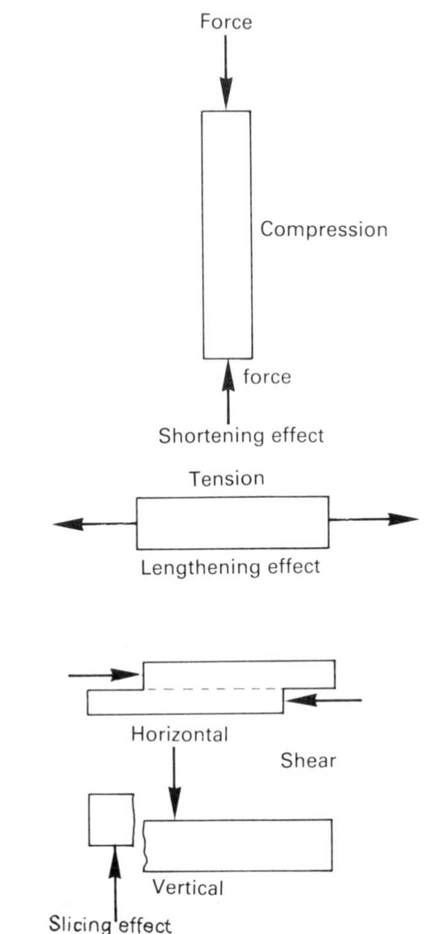

Figure 95 *Types of stress*

Compressive stress Compression causes squeezing and crushing. Its effect is to shorten the member.
Tensile stress Tension tends to pull and stretch. Its effect is to lengthen the member.
Shear stress Shear has a sliding and cutting action and may occur either vertically, horizontally or both.

Structural members when loaded are subjected to a combination of these stresses, as shown in Figure 96.

Horizontal members
Beams, lintels and slabs are subjected to bending, causing an internal combination of compression, tension and shear. The upper half of the member will be in compression, the lower half in tension. Midway between these layers of stress (known as the *neutral stress line*) there will be shear stress as the two layers tend to slide horizonally.

Near the supports vertical shear is encountered. The horizontal and vertical shear stresses work together and produce a tendency to failure along a diagonal line as indicated. This diagonal failure is in fact a result of the combination of the vertical cutting and horizontal sliding action of shear.

Vertical members
Walls and columns are subjected to compression when loaded. However, as loading is rarely concentric they are also subjected to buckling, causing tensile and shear stresses. Excessive buckling in vertical members can occur when they are too tall, too slender or overloaded.

Reinforcement
A beam made of plain concrete under load will bend and progressively crack along its bottom edge due to tension. This cracking reduces the effective depth of the beam, leading to further cracking and eventual collapse.

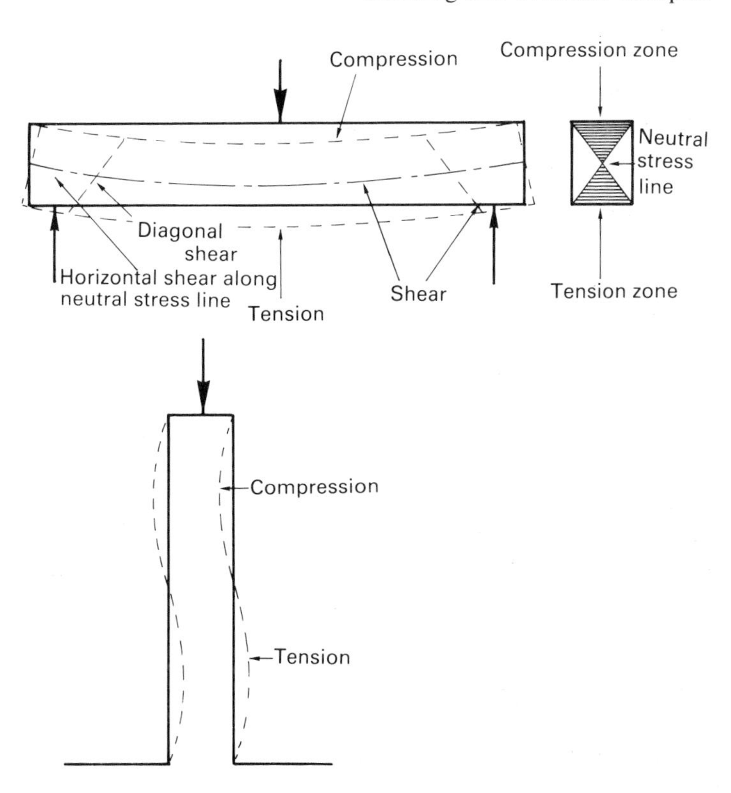

Figure 96 *Stresses in structural members*

In order to resist tensile forces and reduce cracking to a minimum, steel bars are embedded into the concrete. These are positioned near to the bottom edge of beams that are supported at either end (see Figure 97).

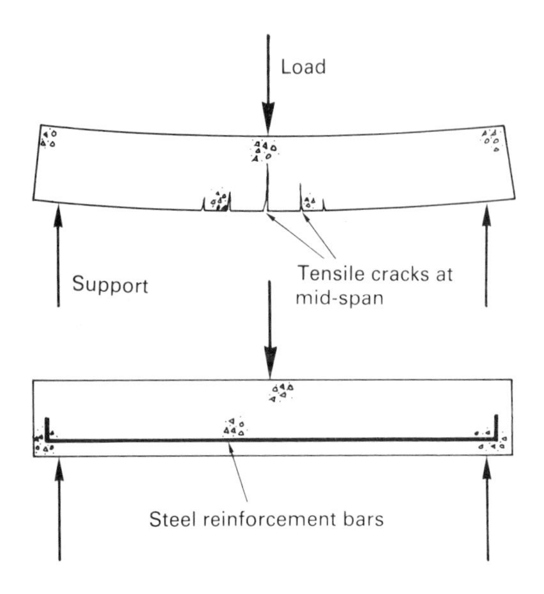

Figure 97 *Purpose of reinforcement*

Where beams have intermediate support the positions of compression and tension stresses are reversed above them. Additional steel bars are therefore located near to the top edge of the beam above intermediate supports, as shown in Figure 98. Plain steel bars are normally hooked or turned up at their ends to provide anchorage in the concrete, preventing any slip between the two materials when under load. Deformed bars have a much greater resistance to slip due to their shape and may be used without hooks.

The combination of horizontal and vertical shear will produce a diagonal line of cracking near to the supports. To resist shear and prevent cracking, either shear bars are put in to act at right angles to the force; or one of the main bars may be cranked up near the supports; or steel stirrups are placed vertically along the beam, spaced closer together near the supports where shear is at a maximum. To prevent misplacement of the stirrups during pouring, top steel bars are included to form a reinforcement cage; or a combination of inclined cranked bars and stirrups may be used. Figure 99 illustrates a typical reinforcement cage for a simple beam.

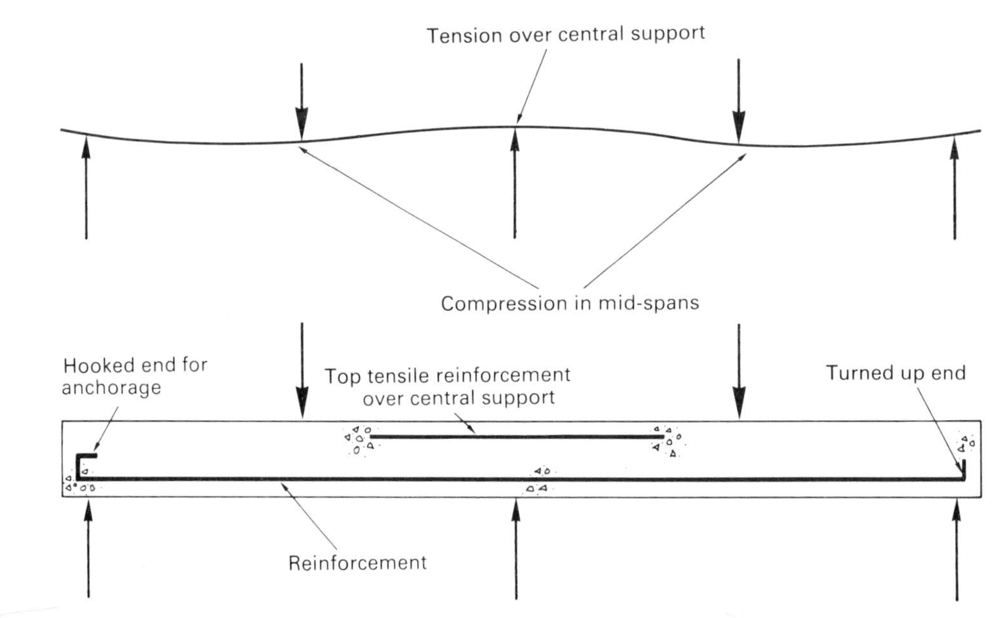

Figure 98 *Beam*

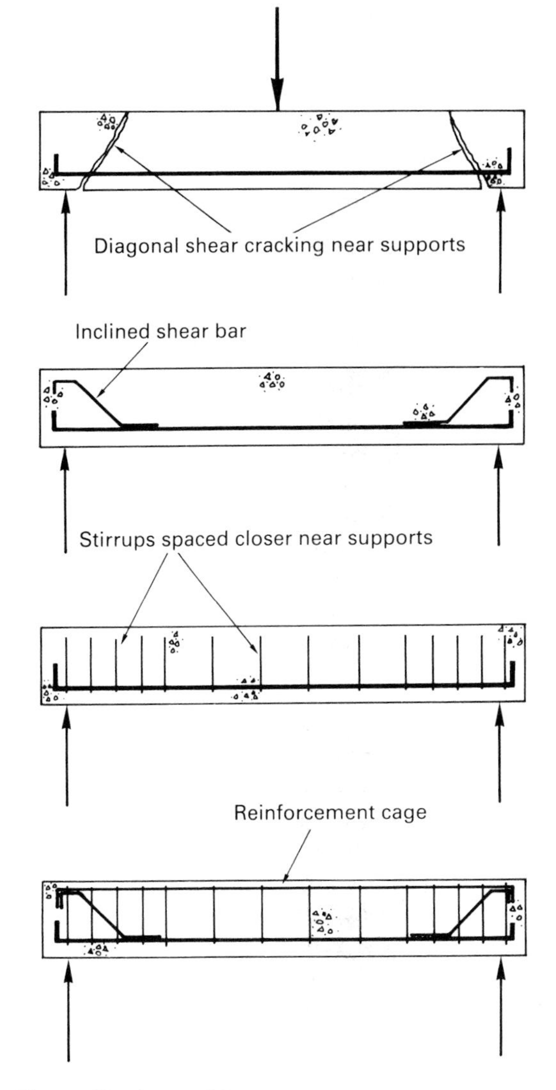

Figure 99 *Typical beam reinforcement*

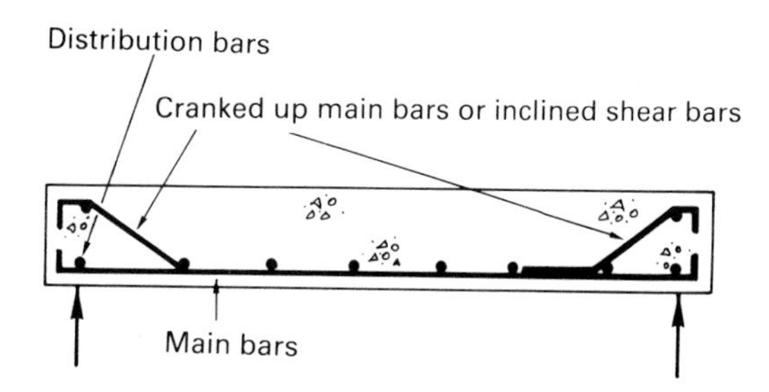

Figure 100 *Floor slab reinforcement*

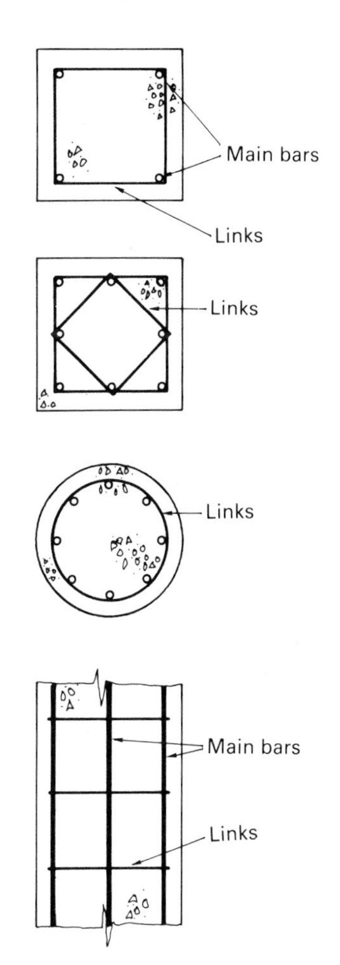

Figure 101 *Column reinforcement*

The reinforcement of suspended floor slabs is similar to that of beams. The main steel will span the shortest direction and be positioned near the soffit. Secondary or distribution steel is wired on top of the main bar in the opposite direction. This resists the tensile forces due to the slab spanning in two directions. Shear stress is controlled by cranking up alternate main bars or by including inclined shear bars (see Figure 100).

Although columns are mainly subjected to compression, vertical reinforcement is included to resist any tensile forces caused through buckling (see Figure 101). The main bars are positioned evenly around the cross-section near to the surface. When loaded the main steel bars themselves will tend to buckle. To overcome this they are tied together in the form of a cage using equally spaced lateral ties or links (also called stirrups).

Walls, like columns, are reinforced owing to the tendency to buckle. The main vertical bars are equally distributed along both wall faces near to the surface. Horizontal distribution bars are tied to the main bars at intervals (see Figure 102).

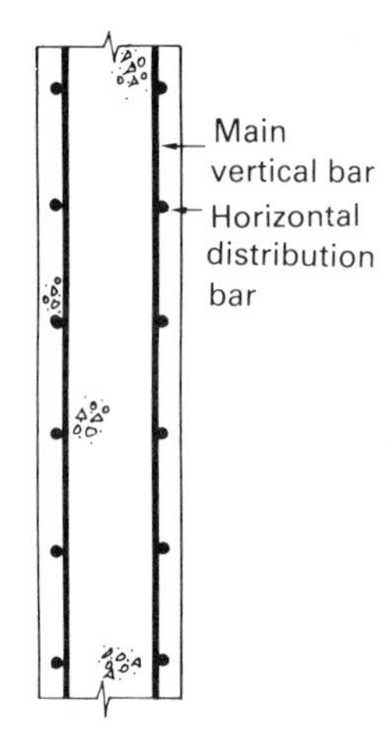

Figure 102 *Wall reinforcement*

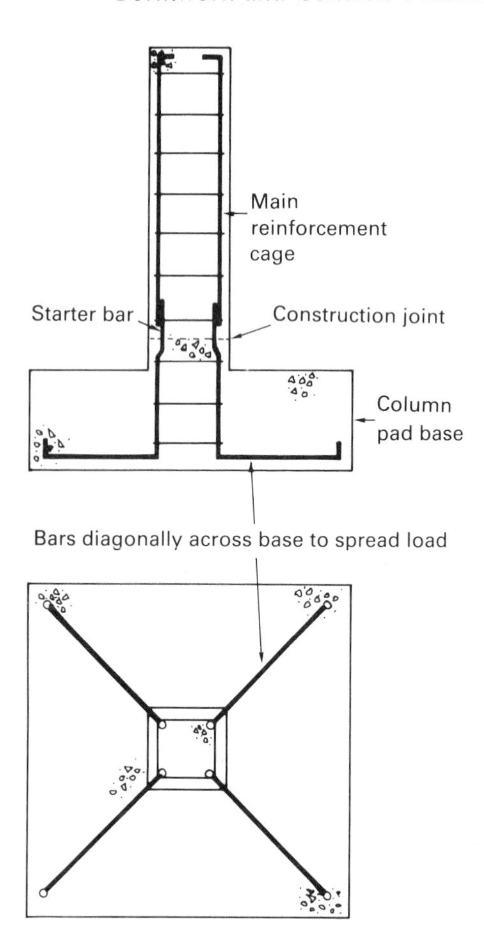

Figure 103 *Column starter bars*

Both column and wall reinforcement is secured to its pad or slab base by tying it to starter bars that have been left protruding from the previous pour, as shown in Figure 103.

Anti-cracking reinforcement may be included near the top surface of slabs and faces of walls to control the development of cracks. The tendency to crack is due to tensile forces set up within the concrete by contraction as cooling takes place after the heat of hydration, and also by shrinkage as the concrete dries out. With reinforcement the cracking is controlled; fine cracks are distributed over the entire surface. Without reinforcement, wide cracks would occur around the middle; this would provide a point of entry for moisture which, apart from other problems, would cause the main reinforcement to lose its corrosion protection (see Figure 104).

Bending schedules

The cutting and bending of the steel reinforcement may be carried out on site by the steel fixer. Alternatively it can be supplied already cut and bent to the required size and shape. Steel layout drawings and bending schedules are used to cut, bend and fabricate the reinforcement in accordance with the structural designer's intentions.

The steel layout drawing and bending schedules for a simple beam are shown in Figure 105. Each bar shape is given a different identifying bar mark number. The diameter and type of bar is included, together with the number of that particular bar mark required in one member and the number of members. The lengths stated are

Figure 104 *Anticrack reinforcement*

the total straight length before bending required in order to form the member. The steel member's finished shape is called its bending shape. These are given British Standard identifying shape codes, e.g. code 20 is a straight bar, code 32 is a hooked end bar, and code 60 is for links or stirrups. Typical standard shape codes are illustrated in Figure 106.

Reinforcement cover

Adequate concrete cover to steel reinforcement (Figure 107) is essential to the durability and fire

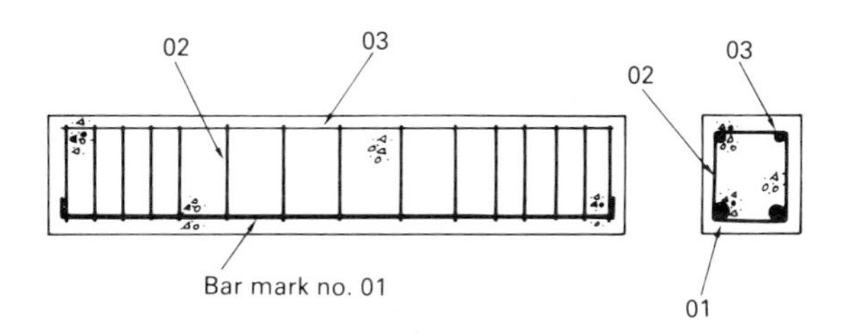

bending schedule

Location Beam	*Type/size (mm)*	*Length (m)*	*Bar mark*	*No. of member*	*No. in member*	*Total*	*Shape code*
B/16	R20	4.10	01	6	2	12	35
	R16	0.80	02	6	15	90	60
	R12	4.00	03	6	2	12	20

Figure 105 *Steel layout and bending schedule*

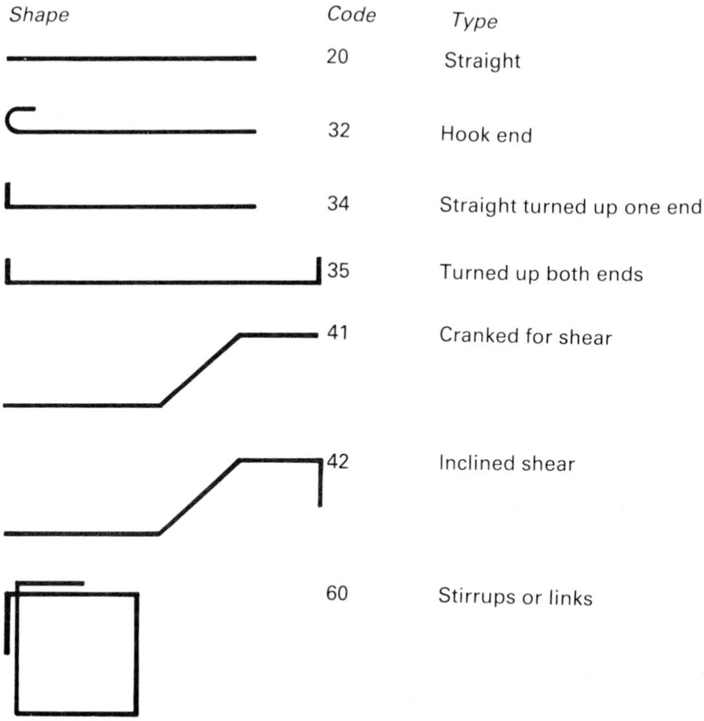

Shape	*Code*	*Type*
	20	Straight
	32	Hook end
	34	Straight turned up one end
	35	Turned up both ends
	41	Cranked for shear
	42	Inclined shear
	60	Stirrups or links

Figure 106 *Typical standard shape codes*

Stirrups or links
Main bar
Cover
Cover
Insufficient cover results in steel expansion and spalling concrete

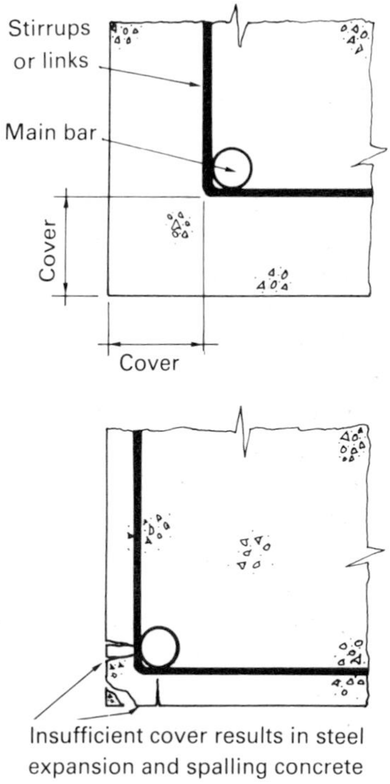

Figure 107 *Concrete cover to reinforcement*

resisting properties of the concrete. Where there is insufficient cover, moisture and air will be able to reach the steel and it will start to corrode.

The layer of corrosion increases the volume of the steel, putting pressure on the concrete and causing it to expand, crack and spall. This leaves the way open for further moisture and frost damage and eventually to structural collapse.

When reinforced concrete with insufficient cover is exposed to very high temperatures in a fire, its steel heats up at an early stage and loses much of its strength, leading to structural collapse.

The actual concrete cover required in any situation is dependent on the quality of concrete, the situation in which it is used, and the period of fire resistance required.

The minimum cover should be at least equal to the diameter of the reinforcement bar and the maximum aggregate size being used. In any case, for normal structural work it should never be less than:

15 mm for internal elements
30 mm for sheltered exterior conditions
40 mm for exposed exterior conditions

Table 19 *Typical fire cover and minimum dimensions for structural concrete elements*

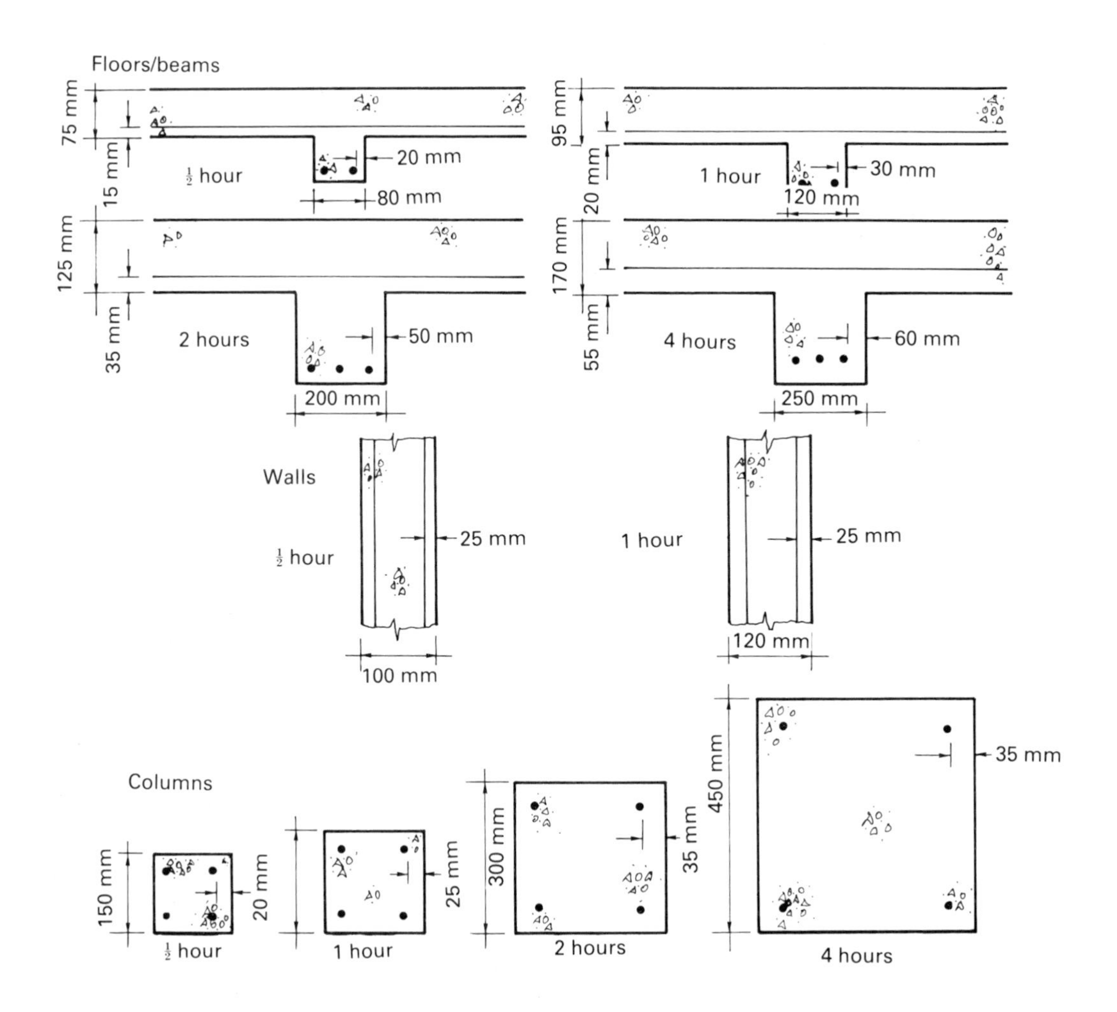

50 mm for air entrained concrete subject to de-icing salts
60 mm for concrete exposed to sea water or moorland water.

The fire resistance of structural elements is measured in terms of integrity and insulation in addition to structural stability. This means that as well as concrete cover to steel reinforcement the element itself must have minimum dimensions according to its intended period of fire resistance. The purpose of this is firstly to prevent cracks opening up which would permit the passage of hot gases and flames through the element, thus destroying its integrity; and secondly to ensure that the temperature on the side of the element remote from the fire does not rise to such a level that would cause materials in the vicinity to spontaneously ignite.

Table 19 shows details of typical fire cover and minimum dimensions for normal density structural concrete elements of various periods of fire resistance.

The correct cover to steel reinforcement can be obtained by the use of spacers that are positioned at intervals between the steel and formwork. Figure 108 illustrates the use of reinforcement spacers.

Prestressed concrete

As concrete is inherently strong in compression, the idea with prestressed concrete is to precompress the concrete member in the area where tensile stresses would develop under normal conditions. Under load the prestressed member acts as if it had a high tensile strength of its own. Tensile cracking will not occur in the member unless the tensile stresses exceed the precompression stresses.

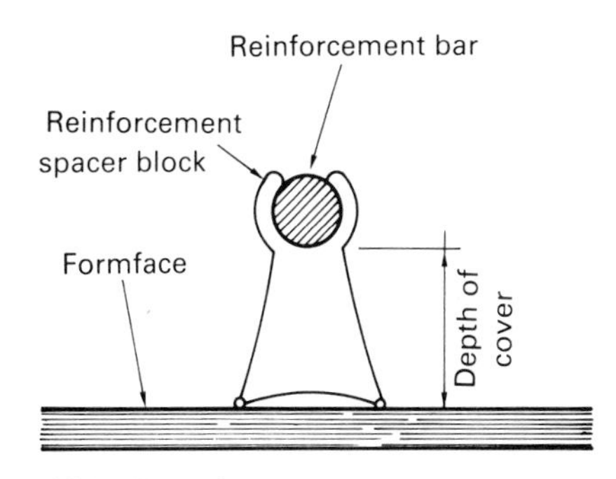

Figure 108 *Use of spacer to maintain cover*

There are two methods by which prestressing may be accomplished, as follows.

Pretensioned concrete

Steel wires or tendons are prestressed between end anchor points and the concrete poured around them (Figure 109). After curing the tensioned steel is released from the anchor points; this has the effect of transferring the stress to the concrete member, putting it into compression. Good bonding between the concrete and steel is essential to the strength of the member.

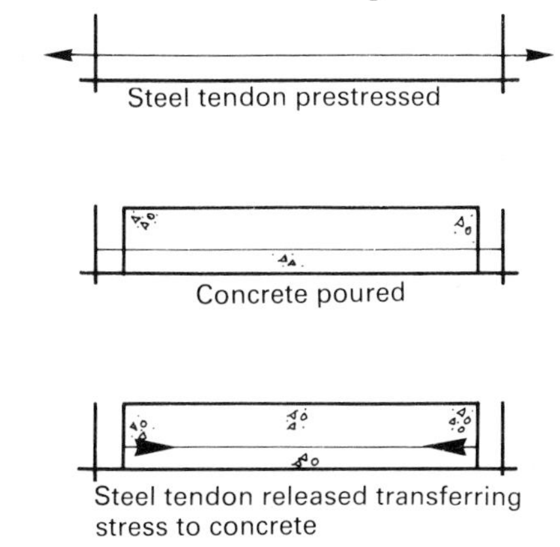

Figure 109 *Pretensioned concrete*

Post-tensioned concrete

This involves casting the concrete member around unstressed steel tendons (Figure 110). After curing the tendons are tensioned using a jacking procedure and held secure with either nuts or wedging devices, thereby compressing the concrete member. The unstressed tendons are prevented from bonding to the concrete by enclosure in steel sheathing. Alternatively ducts may be formed in the concrete member using removable solid or inflatable rubber formers for later tendon insertion. After tensioning and anchoring off, cement paste is pumped around the tendons to provide protection from corrosion and to bond them. Often the tendons are located

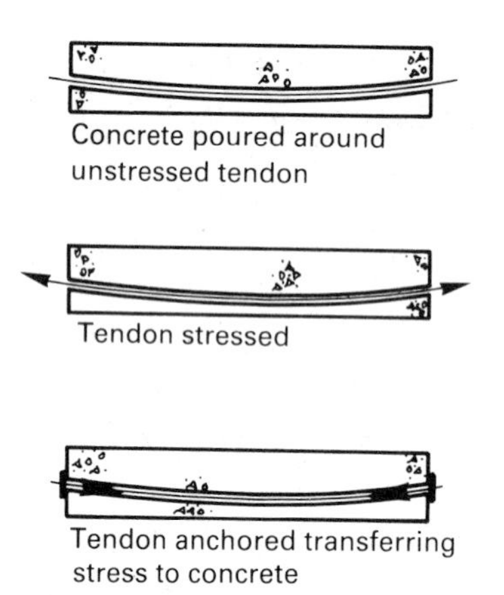

Figure 110 *Post-tensioned concrete*

in a curved profile in order to vary the amount of prestress along the member's length as required.

Surface finishes

In addition to its structural and durability properties, concrete is often selected for use over other materials because of its decorative appearance. As well as plain fair face concrete surfaces and applied decorative finishes there is a wide variety of decorative surface finishes that it is possible either to cast directly into the concrete surface or to tool later.

Exposed aggregate finish

In this type of finish the individual particles of coarse aggregate are exposed on the surface. The main methods employed to achieve this are as follows (see Figure 111):

Early formwork striking followed by water spraying and scrubbing with a stiff brush to remove the surface cement paste. This should be carried out between 12 and 24 hours after pouring.

Surface retarders may be applied to the form face in place of a release agent. These delay the hardening of the cement paste skin at the surface. After striking at about 3 days the skin may be removed by washing and scrubbing.

Acid etching A solution of hydrochloric acid in water is brush applied to the concrete surface. After 5–10 minutes the acid solution and together with it the surface cement paste are removed by clean water spraying and brushing. The degree of aggregate exposure is dependent on the concrete age and the strength of the acid solution.

Grit blasting This involves spraying the surface with a non-metallic grit under pnuematic pressure to erode the surface cement paste and softer aggregate. The depth of exposure and effect achieved will depend on the age of the concrete and the pressure at which the grit hits the surface.

Aggregate transfer In one method the desired exposed aggregate is fixed to the form face prior to pouring using a water based adhesive. As the adhesive bond softens owing to the concrete's moisture, on striking the form face is removed leaving the aggregate embedded into the surface. It then requires only a fine spray and brushing to remove traces of adhesive and grout. In another method the aggregate is carefully placed on a damp fine aggregate bed on the form face. After striking, any loose fine aggregate adhering to the surface can be brushed off. In a third method, coarse aggregate can be individually placed and trowelled into the horizontal surface of fresh concrete to form an exposed aggregate finish.

Tooling This involves chipping the concrete surface with a bush hammer, point chisel, comb chisel or other similar tool. The extent of texture and exposure obtained will vary with the tool and the method of use. Tooling produces a surface that is unlike other exposed aggregates, where aggregates are seen as individual particles. This is because, in addition to removing the surface cement paste, tooling also fractures the aggregate particles, forming a fairly smooth continuous surface.

Whichever method of exposure is used, the removal of the surface cement paste makes the concrete more absorbent. Therefore additional cover should be provided to the reinforcement.

Form lining finishes

This involves lining the form face with a profiled lining material to create a textured raised or indented design. These include (see Figure 112):

Figure 111 *Exposed aggregate finish*

Figure 111 – *continued*

Board marked finish Sawn softwood boards of varying thicknesses are used for the form face. These impart to the concrete surface the board widths, their natural grain pattern and knots etc. in addition to the conversion marks left by circular or band saws. The grain pattern may be highlighted by presoaking the boards in water, or grit blasting their surface to emphasize the difference between the early and late wood of each year's growth. In common with plywood form faces, boards of different ages should not be used in the same form; they have different absorbencies, which will result in distinct colour variation. *Moulded rubber and formed plastic* can be used as form liners to create random or symmetrical designs on the surface and also concrete murals. *Vertical channels* can be formed in the concrete surface by tacking lengths of rope to the form face. On striking the rope is pulled from the concrete at the same time, breaking out the surface and leaving rough cast channels. A more regular appearance is achieved by the use of dovetail shaped timber fillets in place of the rope.

Direct finishes

The two major direct methods of finishing concrete, especially floor slabs, are power trowelling and power grinding.

Power trowelling

Within 2–3 hours of casting the concrete surface will have stiffened sufficiently to enable it to be floated off using a power float. This will compact, smooth and close up any holes in the previously levelled surface. A further 2–3 hours are allowed after floating to permit the surface moisture raised by the float to evaporate. Once the glistening of the surface moisture has disappeared, the surface may be power trowelled. This results in a smooth, dense and durable concrete surface. The timing of this floating and trowelling is critical to the quality of finish achieved, and the whole operation including casting should be carried out during one shift.

The operation can be speeded up by the early removal of excess water from the slab using a vacuum dewatering process. Immediately after

Figure 112 *(a) Form lining finishes moulded rubber; (b) Form lining finishes vertical channels*

compaction and levelling operations are complete the slab surface can be covered with a fine filter sheet, followed by an airtight suction mat. The centre of the suction mat is connected via a flexible transparent hose to the vacuum generator. A vacuum is applied for about 3 to 4 minutes per 25 mm of slab thickness until no further water is seen being sucked along the hose. After removal of the mat and filter the slab is ready for floating.

Power grinding
Grinding of the concrete surface may be carried out between 2 and 7 days after pouring depending on temperature and curing. In any case this early age grinding (as it is sometimes known) must be carried out before the concrete has achieved a high strength.

After the concrete has been compacted and levelled the surface should be smoothed by floating over several times, allowing the raised moisture to evaporate before each subsequent pass. High spots are first removed with the power grinder, followed by coverage of the complete area to remove 1 mm to 2 mm of laitance (surface cement paste) and minor finishing irregularities.

The surface produced will be dependent on the duration of grinding and the fineness of the grinding stones. A surface resembling fine glasspaper is normally the desired effect. Further grinding with finer stones will expose and polish the coarse aggregate, creating a terrazzo effect.

Surface defects

The quality of the finished surface of concrete is dependent on a number of interrelated factors:

1 Strong, accurate, grout tight formwork
2 Suitable form face material
3 Clean form face, treated with correct type and quantity of release agent
4 Impurity-free concrete mix constituents
5 Good concrete mix design
6 Careful transporting and placing of concrete
7 Adequate and uniform compaction of concrete
8 Suitable curing methods
9 Careful formwork striking at the right time.

Ideally the surface of concrete should be cast, with no making good or finishing required. However, even taking extreme care in observing all the previous factors, surface defects or blemishes will occur on occasions. The most common of these are as follows (see Figure 113):
Rust staining on the surface of hardened concrete may normally be attributed to either the presence of impurities in the aggregate, such as iron pyrites, or the failure to clean out from the formwork the snipped ends of the steel fixers' tying wire prior to casting.
Blow holes Small individual cavities on the concrete surface formed by air bubbles, normally associated with either impervious form faces such as steel and glass fibre, or the use of neat oil as release agent or over-application of other types of release agent.
Grout loss Sand textured concrete areas, due to the leakage of cement paste through form face joints.
Honeycombing Coarse stony concrete areas lacking in cement paste and fine aggregate, caused through either excessive joint leakage or poor placing compaction.
Mismatch/misalignment A step, wave or other deviation in a concrete surface, due to poor formwork detailing or erection.
Spalling Damaged or missing areas of surface concrete, due to either frost action, reinforcement corrosion, or striking.
Curtains Hardened grout runs down the face of completed work, due to poor sealing between formwork at completed structure.
Snots Hardened grout runs/fins hanging from the underside of soffits, caused by damaged or poorly jointed decking.
Cracking/crazing Fine network of cracks over concrete surface, due to either shrinkage as a result of cooling and drying out or tensile stress under load.

Many attempts at remedial repair work or finishing to a concrete surface fail to conceal the defect. Often the result is to make it more noticeable. Thus it is far better to take extra care to prevent defects rather than rely on the concrete finisher. Many minor blemishes disappear at the normal viewing distance of 2–3 metres and should be ignored.

Repairs or finishing, where unavoidable, normally take the form of hacking back uneven surfaces, then damping the surface with water and filling with cement paste for fine cracks, voids, blow holes etc., or a cement/fine aggregate mortar for larger repairs. This is followed by rubbing in with a sponge faced float or bagging with hessian sack. Finally a dry cement/fine aggregate mix is rubbed or bagged over the surface to stiffen the material in the cracks and voids. Rubbing or bagging should continue to expose the original concrete surface.

Failure to cure repairs and finished areas will result in cracking and possibly the eventual spalling of the area.

Concrete mixes

Concrete mixes may be specified in one of two ways – by, method or by performance.

Using a method specification, the mix is specified by the proportions of its constituents and the concrete producer uses this as a recipe.

In the case of performance specifications, the structural designer will specify the properties that are required, for example workability and the 28 day cube test strength. The concrete producers will design a mix that meets these requirements using their specialist knowledge.

Nominal mixes

A concrete mix where the proportions of its constituent materials are specified by volume is known as a nominal mix. The specifier will decide what proportions of dry aggregate and cement are to be used. Typical nominal mixes include:

1:3:6 for mass concrete work
1:2:4 for reinforced concrete structures

Most nominal mixes use a 1:2 ratio for the fine to coarse aggregate in order to ensure that there is sufficient fine aggregate to fill the voids between the coarse aggregate particles. The water/cement ratio is not specified, but should be kept to the minimum necessary to give the required workability. This will be determined on site by carrying out mix tests or trials.

Nominal mixes are rarely used today except for some very small non-structural work. Their main disadvantages are the volume measurement inaccuracies due to the bulking of fine aggregate when damp, and the fact that the water/cement ratio is not specified even though it is one of the main factors controlling concrete strength.

Prescribed mixes

In these the constituent materials are specified by their dry mass. BS 5328:1981 describes six prescribed mixes using mix grade numbers, e.g. C7.5P, C10P, C15P, C20P, C25P and C30P. The C stands for compressive strength; the number indicates in N/mm^2 the characteristic strength at 28 days; and the P indicates that it is a prescribed mix. The majority of concrete produced using a prescribed mix will in fact have a compressive strength in excess of the mix's grade number. This is because the characteristic strength of concrete can be defined as the strength value below which not more than 5 per cent of samples fall.

The minimum grade of prescribed mix for reinforced concrete is C25P. This means that its characteristic strength at 28 days will normally be 25 N/mm^2. However, as strength testing is not to be used in judging specification compliance of prescribed mixes, the characteristic strength is not contractually enforceable.

Included in the standard are tables that provide data which can be used to proportion the constituent materials. These tables indicate the amount of aggregate to be used with 100 kg of cement (two bags) after taking concrete grade, maximum aggregate size and required workability into account. Water/cement ratios are not stated in these tables and therefore must be determined by trial mixes.

Concrete mixes produced using the British Standard tables are known as *ordinary* prescribed mixes. *Special* prescribed mixes are used for

Figure 113 *(a) Surface defect rust staining; (b) Surface defects blow holes: (c) Surface defects grout loss; (d) Surface defects honey combing*

Figure 113 – *continued*

concretes not within the scope of the tables, for example lightweight concrete or where admixtures are required.

Yield and batching for prescribed mixes

The following method can be used to determine the yield in cubic metres of a specified mix and also to determine the batching for any particular size of mixer. It assumes that the mass of the concrete will be equal to the mass of its constituents, except for the water. (Although some of the mixing water chemically combines with the cement during hydration, much of it will evaporate. Thus its mass can be ignored.) The density of matured concrete is normally taken to be 2400 kg/m^3.

Example

To determine the yield of a C15P mix and to determine the batching masses for a 750R mixer, using a 20 mm down coarse aggregate, a zone C (coarsely graded) fine aggregate, and medium workability (25–75 mm slump).

Note BS 5328 still specifies fine aggregate in the four outdated zones 1–4. Fine aggregate is now categorized into three zones: F (fine), M (medium) and C (coarse). Until this standard is updated with the new zones, the following approximations can be assumed: zone C can be considered as zone 1, zone M as zone 2 and 3, and zone F as zone 4.

From the BS table use 680 kg of aggregate per 100 kg of cement, and for zone C fine aggregate use 50 per cent of total aggregate.

The yield is as follows:

$$\text{Yield, m}^3 = \frac{\text{mass of constituents}}{\text{density of concrete}}$$

$$= \frac{100 + 680}{2400} = 0.325 \text{ m}^3$$

The quantity of materials to use in a 750 litre (0.750 m^3) mixer is found as follows. The multiplying factor for the mixer is:

$$\text{multiplying factor} = \frac{\text{mixer volume}}{\text{mix yield}}$$

$$= \frac{0.750}{0.325} = 2.308$$

Therefore the quanities required are:

cement = 100 × 2.308 = 231 kg
total aggregate = 680 × 2.308 = 1569 kg

Of the total aggregate, the composition is:

fine aggregate = 785 kg
coarse aggregate = 785 kg

Finally, using a water/cement ratio of say 0.5, the water required is:

water = 231 × 0.5 = 116 litres

Note Adjustment must be made at the time of mixing to take into account the moisture content of aggregates.

Design mixes

These are mixes specified by the structural designer in terms of the concrete's performance, which is normally the workability and the 28 day compressive cube strength. The concrete producer is then responsible for designing and supplying a mix that will meet the specifier's stated requirements.

In design mixes the cube testing for compressive strength will normally be the major factor in determining specification contract compliance. Where compressive strength is specified, concrete shall be assumed to comply if:

1. The average strength of four consecutive tests exceeds the characteristic strength by at least 3 N/mm^2 for C20 grade concretes and above, or 2 N/mm^2 for lower grades; and
2. The strength of any one test result does not fall below the characteristic strength by more than 3 N/mm^2 for C20 grade concretes and above, or 2 N/mm^2 for lower grades.

BS 5328 lists the following 15 grades for design mixes: C2.5, C5, C7.5, C10, C12.5, C15, C20, C25, C30, C35, C40, C45, C50, C55, C60. The C indicates that the number is the characteristic compressive strength in N/mm^2.

In addition the standard states three grades for flexural (bending) strength concrete (F3, F4 and F5) and three grades for indirect tensile strength concrete (IT2, IT2.5 and IT3). Again in each case the grade numbers indicated are the characteristic strengths in N/mm^2 for 28 days.

Designers are of course free to use their experience and knowledge in order to design other grades of concrete outside the BS range of grades.

Choice of mix

The decision on what type of mix to specify is dependent on a number of factors:

Nominal mixes are not generally recommended, although they may still be useful for the small builder without mass batching facilities when undertaking small non-structural work.

Prescribed mixes are more likely to be specified for small or intermittent quantities of concrete where strength testing facilities are not available.

Design mixes are specified for contracts using large amounts of concrete, of either a structural or a non-structural nature. These produce a high quality concrete that is more economical and consistent than either nominal or prescribed mixes.

Self-assessment questions

Question	*Your answer*
1 Define the term *hydration* with regard to cement.	
2 Determine the water/cement ratio for a mix containing 350 kg of cement and 180 litres of water.	
3 Define what is meant by the *workability* of concrete.	
4 Name *three* factors that affect the workability of a concrete mix, and state the effect of *each*.	

5 Name *three* factors that affect the 28 day compressive strength of concrete, and state the effect of each.

6 Briefly explain the characteristics of concrete that affect its thermal, fire and sound insulation properties.

7 Describe using a practical example the *three* types of stress to which structural members may be subjected.

8 Sketch a typical reinforcement detail for a concrete beam supported at either end and midspan by columns.

9 Describe what is meant by cover to steel reinforcement.

10 Describe *three* methods of producing exposed aggregate concrete finishes.

11 Define and state the probable cause of the following surface detects:
(a) Blow holes
(b) Honeycombing
(c) Spalling.

12 Distinguish between concrete mixes specified by method and performance, and state an example of *each*.

13 The presence of pyrites in an aggregate could result in:
(a) An increase in strength
(b) Rust staining
(c) Low workability
(d) High durability.

a [ ]	b [ ]	c [ ]	d [ ]

14 Concrete members that are cast around stretched steel wires are described as:
(a) Reinforced
(b) Post-tensioned
(c) Pretensioned
(d) Stressed.

a [ ]	b [ ]	c [ ]	d [ ]

Chapter 5

Concrete materials and testing

Materials for concrete

As determined in Chapter 4, concrete is a composite material made from a mixture of cement, aggregate and water with the addition in certain situations of chemical additives or admixtures. It is possible by careful selection of types of cement, aggregate, admixtures and handling methods to produce a concrete having properties specific to a particular situation.

When selecting materials and determining mix proportions for concrete, the following factors may require consideration:

Strength
Density
Cost
Surface finish/colour
Thermal insulation
Chemical resistance
Waterproofing
Shrinkage and expansion.

Therefore the actual materials and the proportion to be used in a particular concrete will be determined by the specific properties required. For example, dense concretes have a fairly high strength, a hard surface, and good resistance to chemical attack, water and sound penetration. On the other hand, lightweight concretes have a low density, a fairly soft surface, excellent thermal insulation properties, and high shrinkage and expansion rates.

Cement

Cement is the most important and expensive constituent of concrete. It is manufactured from chalk or limestone and clay which are ground into a powder, mixed together and fired at a very high temperature in a kiln, causing a chemical reaction resulting in the formation of a clinker. After leaving the kiln the clinker is ground to a fine powder, to which is added a small amount of gypsum to control the rate of settling when the cement is subsequently mixed with water.

By modification of the initial materials and/or addition of other materials during the manufacturing process, it is possible to produce a wide range of cements with differing properties to suit a wide range of end uses. Table 20 shows some typical uses for different cements.

Table 20

Type of cement	*Typical use*
Ordinary Portland cement	Good general purpose cement
Rapid hardening Portland cement	In cold weather or when time is important
Sulphate resisting Portland cement	Where sulphates are present in the ground. In mass concrete pours
Low heat Portland cement	Massive concrete pours with high risk of cracking due to excessive thermal stresses
Portland blast furnace cement	Mass pours and warm weather working
White Portland cement	Decorative pre-cast panels, copings, pavings etc.
Hydrophobic cement	In damp, poor storage conditions

Ordinary Portland cement (OPC)
This is the most widely used cement in concrete. It gains its name from the fact that its hardened appearance closely resembles that of Portland stone. It is suitable for general concreting work; gains strength at a steady rate; will resist temperatures of up to 150°C without significant deterioration or loss of strength; but is susceptible to attack by acids and sulphates which may be present in certain soils and ground waters.

Rapid hardening Portland cement (RHPC)
This is similar to OPC except that it has been more finely ground to produce a more rapid rate of hardening. This occurs since a greater surface of area is exposed to water. Typically the strengths of RHPC concrete at 1 and 3 days are similar to the strengths of OPC concrete at 3 and 7 days respectively. By 28 days there is little difference in the strength of the two, and ultimately it will be identical. The early strength development enables formwork to be struck at an early stage. Associated with early strength development is a high rate of heat development, which is useful in combating the effects of low winter temperatures. However, in normal conditions there is a greater risk of thermal cracking and thus RHPC concrete is not usually considered suitable for mass concrete work.

Sulphate resisting Portland cement (SRPC)
This cement is formulated to give increased resistance to attack by sulphates present in the soil, ground water and sea water than either OPC or RHPC. This is achieved by reducing the amount of aluminate (the mineral most affected by sulphate salts) present in the raw mix. Sulphate resisting cement produces less heat than other Portland cements, so in addition to its main use in grounds where sulphates are present it can also be used to advantage in mass concrete pours.

Low heat Portland cement (LHPC)
This has a lower lime content than OPC, which results in less early heat and thus a low rate of strength development. Its 7 day strength is similar to the 3 day strength of OPC. The low heat properties of this cement make it ideal for use on really massive concrete pours where there is a high risk of cracking due to excessive thermal stresses.

Portland blast furnace cement (PBFC)
This cement is a blend of OPC and finely ground granulated blast furnace slag. The slag slows the early development of heat and thus strength, but it does make it more resistant to chemical attack particularly by sea water. This slow development makes it suitable for mass pours and warm weather working. In common with LHPC, PBFC is not suitable for use in cold weather and where the early release of formwork is required.

White Portland cement (WPC)
This requires careful selection of the raw materials to limit the iron content that gives Portland cement its characteristic grey colour. White cement is mainly used for decorative precast panels, copings, pavings and other objects. Apart from its colour, its other properties are similar to OPC.

Hydrophobic cement
This cement is manufactured for use in damp, poor storage conditions, where other cements may start hydrating prematurely. A water repellent coating is formed around each cement particle during manufacture by intergrinding the clinker with a hydrophobic substance (a substance that does not easily mix with water). The abrasive action of the mixing process removes this coating, leaving a cement with the same properties as OPC.

High alumina cement (HAC)
This is not a Portland cement. It consists of limestone and bauxite (an aluminium ore) fused together in a furnace, crushed and finely ground to form a cement that is darker grey in colour than OPC. Strength development is very rapid. HAC achieves the majority of its strength within 24 hours; its 1 day strength is equivalent to the 28 day strength of OPC. Typically the final strength of HAC is about twice that of OPC. HAC is

highly resistant to sulphate and acid attack and is also suitable for use in high heat conditions. However, it has been found that when hardened HAC concrete is exposed to fairly hot (above 25°C) and moist conditions, a chemical reaction occurs known as conversion. During conversion the hydration products form into a different structure having as little as 30 per cent of its original strength. Building failures have been associated with the structural use of HAC concrete. Therefore it is no longer recommended for structural work.

Contamination with Portland cement causes a *flash set* (very rapid setting) when water is added. Thus HAC must be stored separately from other cements. Careful cleaning of all plant items is also essential when alternating between HAC and Portland based concretes.

Aggregate

Aggregates are the sands, gravels and crushed rocks which are added to cement in order to produce a concrete with specific properties. Aggregates form the bulk of the material in a concrete mix and may be added for any one or a combination of the following reasons:

1 Aggregates are considerably cheaper than cement.
2 Aggregates can increase durability, as suitable aggregates have a greater resistance to weathering and abrasion than does a cement paste.
3 Aggregates are better at accommodating shrinkage than is a cement paste.
4 Aggregates can alter the density of the mix, e.g. lightweight, normal and heavy.
5 Aggregates can be added to give a mix special characteristics, e.g. fire resistance, thermal and sound insulation.
6 Aggregates can be used to achieve various surface finishes and colours.

Characteristics of aggregates

In order to produce good quality concrete the aggregates should possess the following desirable properties.

Durability
Aggregates must be hard and not contain materials which are liable to any form of shrinkage, swelling or decomposing when exposed. In addition aggregates for reinforced work must not contain any materials which will attack the steel reinforcement. For example: clay softens or swells, forming weak pockets in the concrete; coal swells; iron pyrites decompose and often cause unsightly rust stains on the finished concrete; and salts and clay affect the setting and hardening of the cement paste. in addition, salt can cause corrosion in the steel reinforcement and efflorescence on the finished concrete.

Cleanliness
It is essential that aggregates are free from any dust and clay coatings or other impurities. These can prevent the aggregate being satisfactorily coated with the cement paste, thereby restricting bonding and producing concrete of a lower strength. Sands and gravels may be washed by the supplier prior to delivery on site to ensure cleanliness.

Types of aggregate

Aggregates may be classified by:

Origin as natural or manufactured
Particle size as fine or coarse
Density as lightweight, normal and heavy
Particle shape and texture
Grading

Origin

Aggregate origins are shown in Figure 114:

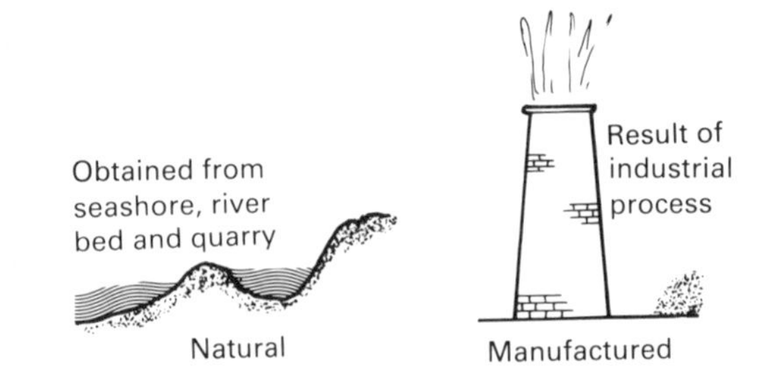

Figure 114 *Origin of aggregate*

Natural aggregate is defined as an aggregate obtained from mineral sources. These include sand, gravel and crushed rock which is obtained from river beds, sea shores and quarries.
Manufactured aggregate is defined as an aggregate resulting from an industrial process usually involving some form of heating, such as crushed brick, blast furnace slag and numerous lightweight expanded aggregates.

Particle size
Aggregates normally contain a range of particle sizes (Figure 115). The three sizes that are normally specified are defined as follows:
Fine aggregate A natural or manufactured aggregate having a particle size that will be mainly retained on a 75 μm square sieve and passing a 5 mm square sieve. (A micrometre or micron (μm) is 1/1000 of a millimetre.)
Coarse aggregate A natural or manufactured aggregate having particles which are mainly retained on a 5 mm square sieve.
All-in aggregate A natural or manufactured aggregate consisting of a mixture of fine and coarse aggregates.

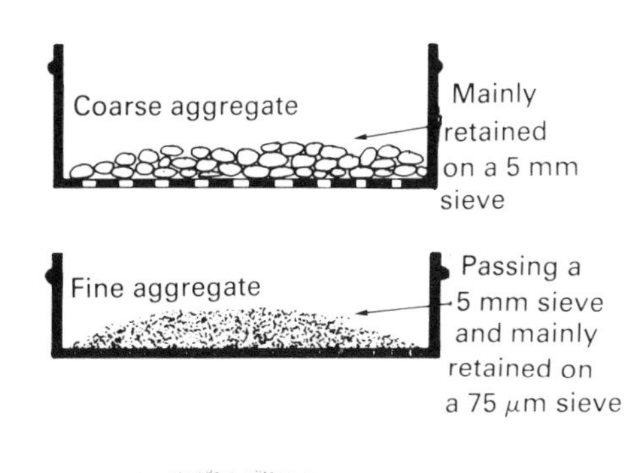

Figure 115 *Aggregate particle size*

Density
Aggregates with a bulk density of up to 1200 kg/m^3 for fine and up to 1000 kg/m^3 for coarse are termed *lightweight aggregates*. Those with bulk densities of 4000 kg/m^3 and over are termed *heavy aggregates*. Aggregates with bulk densities between these two are known as *normal density aggregates*, with bulk densities in the range 1200 to 2000 kg/m^3 being the most common. (See Figure 116.)

Lightweight aggregates
The low density is a result of a cellular porous structure. They are lower in strength than normal density aggregates but have the advantage of lowering the dead weight of a structure and improving its thermal insulation and fire resisting properties. The main types of lightweight aggregate are:

Clinker The coarse residue from coal burning furnaces. Not commonly available today owing to the decline in coal fired industrial furnaces. Because of its high sulphur content it should not be used in contact with steel as corrosion will result.
Foamed slag Molten blast furnace slag is rapidly cooled in water under controlled conditions, causing it to bloat and form a hard sponge – like material similar to natural pumice.
Sintered pulverized fuel ash (PFA) The fine ash residue remaining after burning pulverized fuel

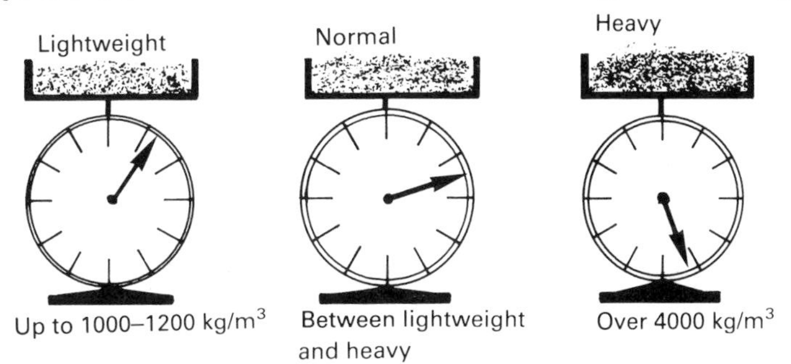

Figure 116 *Aggregate density*

(powdered coal) in power stations is recovered from the flue gases and mixed with water to form pellets. These are subjected to a heat treatment known as sintering which compresses the pellets into a solid body but at a temperature below melting point. After crushing and grading, sintered PFA is suitable for use as either coarse or fine aggregate.
Expanded clay, shale, perlite and exfoliated vermiculite These are all natural materials that have been expanded into a cellular structure using heat treatment. On cooling the cellular structure is retained. Certain clays and shales, when heated almost to softening point, expand due to the generation of gases within the material. Perlite is a glassy volcanic rock that contains moisture. After crushing the rock is rapidly heated to a very high temperature, causing the formation of air bubbles within the material. Vermiculite is a flaky, layered mineral. On heating the layers exfoliate (open out) forming small concertina-like particles. Vermiculite greatly improves concrete's thermal insulation and fire resisting properties.
Pumice is a natural highly porous hard sponge-like material of volcanic origin, formed by the expansion of gases in molten lava and contact with moisture. It has excellent thermal insulation properties. However, it is expensive to import and is thus only occasionally used.

Normal density aggregates
The majority of these are natural gravels, sands and crushed rocks. *Gravel* is a coarse aggregate resulting from the natural disintegration of rock. *Sand* is a fine aggregate; the term should only strictly speaking be used to described a fine aggregate resulting from the natural disintegration of rock. Other fine aggregates produced by crushing rock or gravel should be termed *crushed rock fines* and *crushed gravel fines* respectively.

These natural aggregates are all derived from rock which can be classified into three groups:

Igneous rock was formed by the original solidification of the earth's molten material. It is extremely hard wearing and strong, especially in compression. Granite is the main example.
Sedimentary rock was formed by a settling process in which rock and organic particles were deposited in successive layers on the sea and river beds. These layers subsequently compacted to form a solid material. Limestone and sandstone are the main examples.
Metamorphic rock was formed from older rocks that had been subjected to very high pressures and temperatures, causing a structural change to take place. The two main examples are marble formed from limestone, and slate formed from clay.

Igneous granites and sedimentary limestones are used for the majority of aggregates. Sedimentary sandstones are often too soft for concrete aggregate. Metamorphic rocks are normally unsuitable for aggregates, although they are used to provide decorative surface finishes.

The two main normal density manufactured aggregates used for concrete are as follows:

Blast furnace slag is the waste material drawn off during the manufacture of steel. After cooling and crushing it is suitable for use as a coarse aggregate in concrete.
Crushed brick Clean good quality crushed brick may be used as a coarse aggregate for mass concrete pours. Brickwork obtained from demolition may be economical, but it is essential that thorough cleaning takes place to ensure that it is free from adhering mortar, plaster, dust, timber, soluble sulphates etc. Concrete produced using crushed brick aggregate has a high fire resistance.

Heavy aggregates
These are used to produce a high density concrete suitable for use as radioactive screens and in situations where a high degree of sound insulation is required. The main aggregates used for heavy concrete are iron ores, lead ores and steel.

Particle shape and texture
The particle shape of aggregates has an effect on the workability of a mix, the ultimate strength of the concrete and also the percentage of voids (Figure 117).

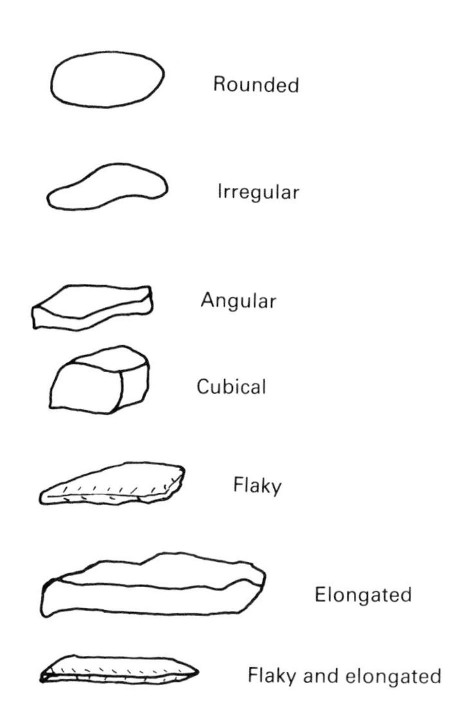

Figure 117 *Aggregate particle shape*

Coarse aggregates may be classified by particle as follows:

Rounded A coarse aggregate the particles of which have rounded surfaces.
Irregular A coarse aggregate the particles of which have an irregular shape and rounded edges.
Angular A coarse aggregate the particles of which have sharp edges.
Cubical A coarse aggregate the particles of which form an approximate cube.
Flaky A coarse aggregate the particles of which are fairly thin in relation to their length and width.
Elongated A coarse aggregate the particles of which are fairly long in relation to their width and thickness.
Flaky and elongated A coarse aggregate the particles of which are both long and thin.

Rounded edge aggregates produce a mix that has a high workability as the edges easily slip over each other, although this also has the effect of reducing strength. Angular and cubical aggregates used in concrete produce a mix with a low workability but a high strength. This is because the particles interlock and provide support and bearing for each other. The interlocking of aggregates also reduces the amount of voids. Aggregates with a large proportion of flaky or elongated particles are normally unsuitable for concrete.

Fine aggregates may be classified as either sharp or soft. Sharp fine aggregate has angular shaped particles and is used for concrete, whereas soft fine aggregate has rounded particles and is used in brick laying mortars and plaster.

The surface texture of aggregates is an important consideration as this affects the bonding of the cement paste. These may range from glassy smooth through to rough honeycomb. In general the rougher surface textures produce a good mechanical bond, whereas the smoother textures provide good cohesion.

Grading

Grading is the term used to describe the proportion of different particle sizes in an aggregate. This has a bearing on the workability of a mix. An aggregate containing a high proportion of large particles is termed *coarsely graded* and one containing a high proportion of small particles is *finely graded*. An aggregate that falls between these two is classified as a *medium grade*.

The maximum coarse aggregate size for unreinforced mass concrete pours is normally 40 mm; for reinforced concrete this is normally reduced to 20 mm. This may require further reduction depending on the size of the section being cast and the amount of steel reinforcement. In general the largest aggregate particle should not be more than a quarter of the smallest section to be cast, and should be 5–10 mm smaller than the minimum gap between adjacent steel reinforcing bars.

To determine the particle size distribution in an aggregate it is necessary to shake a sample through a series (nest) of sieves of standard mesh sizes. The result is expressed as a percentage of

Sieve size	Nest of sieves	Mass retained (g)	Mass passing (g)	Percentage passing
37.5 mm		0	1500	100
20.0 mm		225	1275	85
0.0 mm		75	1200	80
5.0 mm		150	1050	70
2.36 mm		150	900	60
1.18 mm		75	825	55
600 μm		150	675	45
300 μm		225	450	30
150 μm		300	150	10
Pan		150	0	0

Figure 118 *Sieve analysis*

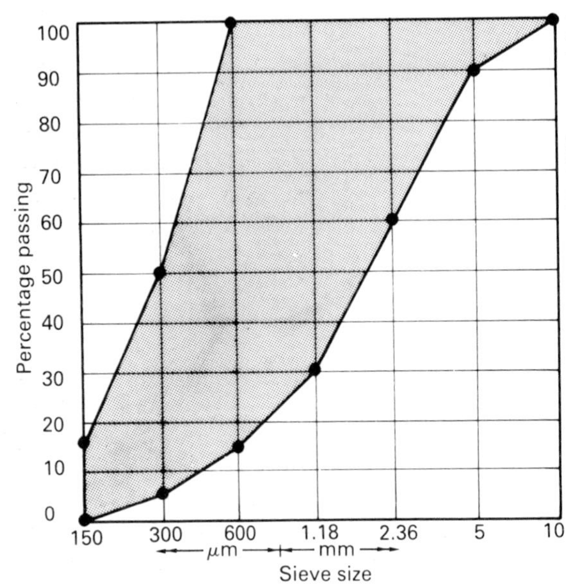

Sieve size	Percentage passing
10.00 mm	100
5.0 mm	89–100
2.36 mm	60–100
1.18 mm	30–100
600 μm	15–100
300 μm	5–70
150 μm	0–15*

* Can be increased to 20% for crushed rock fines.

Figure 119 *Grading limits for fine aggregate*

the sample mass passing retained on each sieve, commencing with the largest (see Figure 118).

It is possible to produce good quality concrete using any fine aggregate that falls within the overall grading limits illustrated by Figure 119. In situations which require greater control it is possible to specify a fine aggregate by its zone classification i.e. zone F (fine), zone M (medium) and zone C (coarse). These are illustrated in Figure 120.

Coarse aggregates are classified as graded aggregates which contain a range of particle sizes, or as a single sized aggregate which mainly consists of one particle size. Graded coarse aggregates are often referred to by their *maximum size and down*; for example, 20 mm and down means that the aggregate contains 20 mm particles and suitable proportions of smaller particles. Figure 121 illustrates the grading limits for 40 mm to 5 mm and 20 mm to 5 mm graded aggregates, and 40 mm and 20 mm single sized aggregates.

The surface area for a given amount of aggregate increases as the particle size of the aggregate

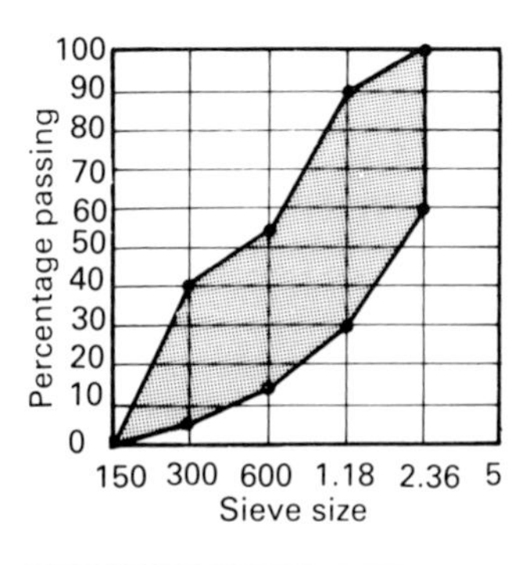

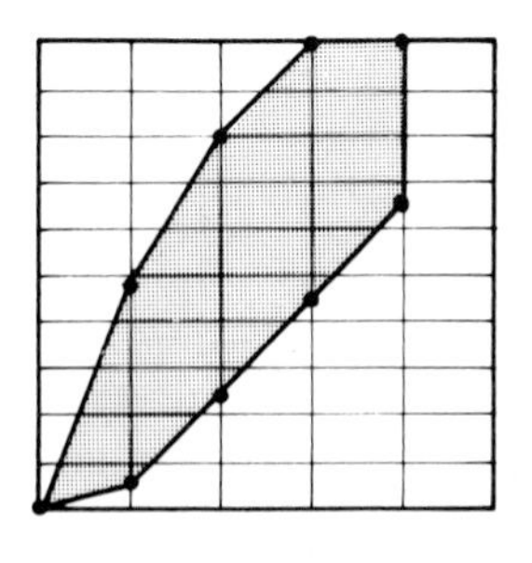

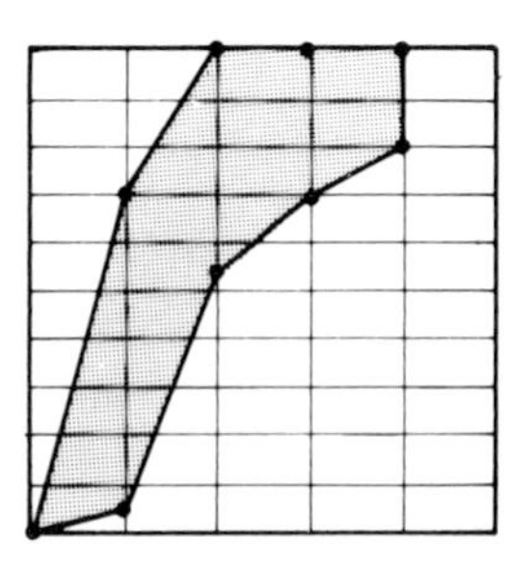

Sieve size	*Zone C percentage passing*	*Zone M percentage passing*	*Zone F percentage passing*
5.0 mm	—	—	—
2.36 mm	60–100	65–100	80–100
1.18 mm	30–90	45–100	70–100
600 μm	15–54	25–80	55–100
300 μm	5–40	5–48	5–70
150 μm	—	—	—

Figure 120 *Zone classification of fine aggregates*

Sieve size	*Percentage passing*			
	Graded aggregate		*Single-sized aggregate*	
	40–5 mm	*20–5 mm*	*40 mm*	*20 mm*
50.0 mm	100	—	100	—
37.5 mm	90–100	100	85–100	100
20.0 mm	35–70	90–100	0–25	85–100
14.0 mm	—	—	—	—
10.0 mm	10–40	30–60	0–5	0–25
5.0 mm	0–5	0–10	—	0–5
2.36 mm	—	—	—	—

Figure 121 *Grading limits for coarse aggregate*

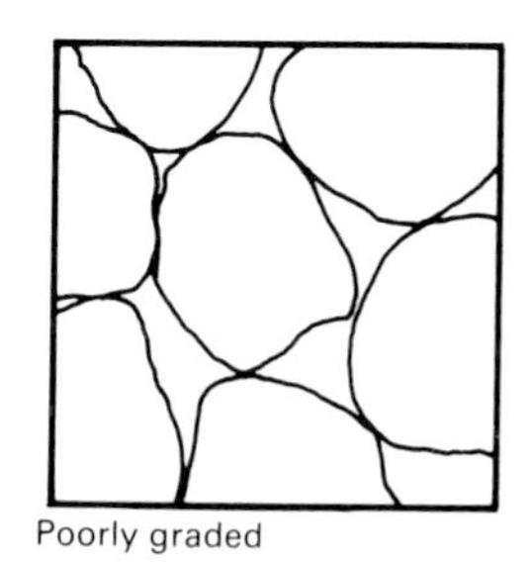

Poorly graded

All particles about the same size

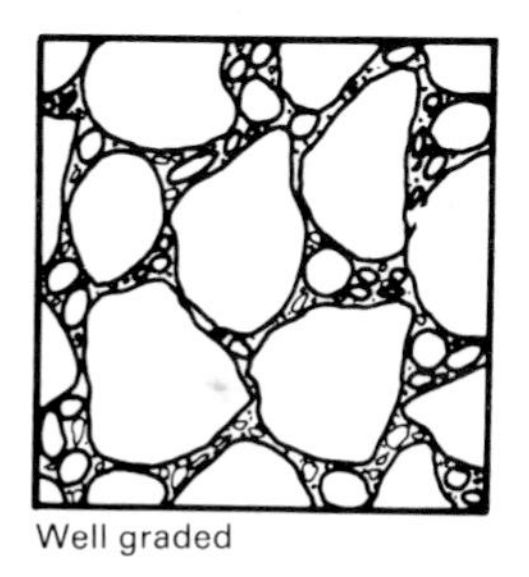

Well graded

A range of different particles sizes

Figure 122 *Poor and well graded aggregate*

decreases. Therefore mixes using a finely graded aggregate have a lower workability for a given water/cement ratio (amount of cement paste) than do medium or coarsely graded aggregate mixes.

Aggregate grading is also one of the factors that affect the proportion of voids in a concrete mix. In order to minimize voids it is essential that a well graded aggregate is used containing a range of particle sizes from large to small. The intention is that the smaller particles will occupy the spaces between the larger ones (see Figure 122).

Water

Water is required in a concrete mix to convert the cement to a paste, hydrate it and provide workability. In general water that is supplied through the mains as fit for drinking is normally considered suitable for concrete production.

In remote locations it may be necessary either to transport mains water to the site by tanker, or to draw water locally from a stream or river. Water drawn from the latter may be contaminated with industrial waste and vegetable matter which can affect hydration of the cement and the long term durability of the concrete. If good quality concrete is required and the water is suspect it should be sent to a laboratory for analysis before use.

Admixtures

An admixture is a material that is added to the basic constituents of a concrete mix (cement, aggregate, water) in order to alter one or more of its properties in its fresh or hardened state.

The use of admixtures must be strictly controlled. Dosages should be in accordance with the manufacturer's instructions. Admixtures are supplied as either liquids or a powder which are added to the mixing water. Powder admixtures normally require to be dissolved in water before use.

An attempt to alter multiple properties of concrete by the addition of two or more admixtures to one mix is not normally recommended, as the results can be unpredictable. The manufacturers should be consulted if a combination of admixtures is to be used.

There is a wide variety of admixtures available for use. They can be classified under the following main groups.

Accelerators

These are chemicals that speed up the rate of hydration, thus increasing the rates of setting, gain in strength and heat generation. They are mainly justified for use in cold weather concreting and/or where earlier striking of formwork is required.

Chloride based accelerators must not be used in reinforced or prestressed concrete as they cause corrosion in steel. In addition chloride also reduces the resistance of concrete to sulphate attack and therefore should not be used for concrete in contact with sulphates.

Retarders

These are chemicals that coat each cement particle to slow down the rate of hydration. The main reasons for using retarders include hot weather concreting and situations where the mixed concrete has to be transported some distance.

Water-reducing admixtures

These are also termed *workability aids* or *plasticizers*. They lower the surface tension of the mixing water, which improves the dispersion of the cement particles and lubricates the cement paste. They are available as normal, accelerating and retarding, the latter two combining the function of both an accelerator (retarder) and a normal water-reducing admixture. They are mainly used either to improve compressive strength by reducing the water content but maintaining workability, or to improve workability without the addition of extra water. In addition they also improve the cohesiveness of the mix, reducing the likelihood of segregation in highly workable mixes.

Super plasticizers

These are a fairly recent admixture development. They are able to dramatically increase the workability of a mix without causing loss of strength or segregation. The chemical is absorbed into the cement particles, causing them to become negatively charged and thus repel each other. Concrete mixes with a slump of 75 mm can achieve a slump in excess of 200 mm, producing a flowing concrete suitable for densely reinforced members using the minimum of vibration. This flowing property also speeds up the casting of large horizontal areas and underwater work using tremie pipes. In addition super plasticizers can be used to produce high strength concrete, in situations where increased workability is not required, by reducing the mixing water by up to 30 per cent.

Owing to the fairly short effective life of super plasticizers (up to 1 hour) concrete should be placed with the minimum of delay. Ready mix trucks should add super plasticizers at the site just prior to discharge.

Air-entraining agents

These reduce the surface tension of the mixing water and introduce millions of spherical evenly distributed bubbles (about 5 per cent volume) which are retained in the hardened concrete. Although this reduces strength, it increases workability and reduces the risk of segregation.

The main reason for using air-entraining agents is to increase saturated hardened concrete's resistance to frost damage and de-icing salts, making them particularly suitable for the top surface of concrete road slabs, drives etc. They work in two ways. Firstly, as the air bubbles are compressible they are able to accommodate the expansion of moisture on freezing. Secondly, the bubbles reduce the capillarity of the concrete, thus reducing the amount of moisture and de-icing salt solution being drawn into the surface.

Steel

Steel reinforcement may consist of mild steel bars, either plain or deformed; twisted or deformed high yield bars; and welded or interwoven mesh fabrics. These are illustrated in Figure 123.

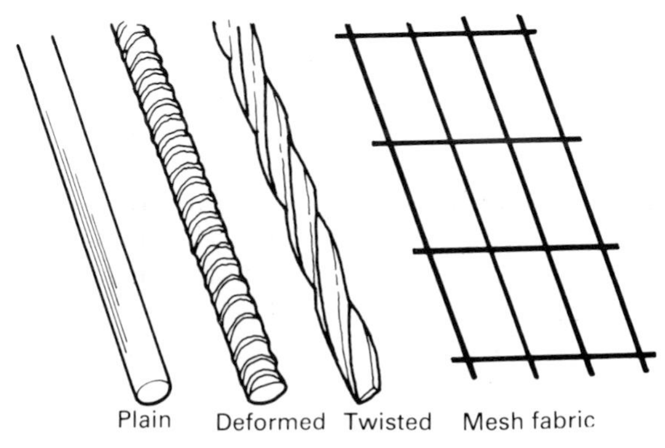

Figure 123 *Steel reinforcement*

It is the normal practice with plain mild steel bars to hook or bend up the ends of the bars in order to provide anchorage and prevent any slip between the steel and concrete when a member is loaded. The use of deformed bars offers a greater resistance to any slip, and can do away with the need to use hooked or bent ends. High yield or high strength steel is likewise deformed to provide anchorage, thus preventing slip.

Most bars are supplied in 12 m lengths with preferred diameters of 8, 10, 12, 16, 20, 25, 32 and 40 mm. There are also smaller 6 mm bars and larger 50 mm bars available should the design require them. Bars are normally purchased by weight and specified by type and size. To avoid confusion on site it is advisable to have different types and sizes of steel either tagged or colour coded with a splash of paint on their ends.

Steel fabrics are available having grid sizes of 200 mm × 200 mm, 100 mm × 200 mm and 100 mm × 400 mm in sheet sizes of 2.4 m × 4.8 m or rolls 2.4 m wide and either 48 m or 72 m in length. Again these are normally sold by weight to the nearest sheet or roll.

Storage of steel

All reinforcement should be stored flat and level, preferably on concrete or timber sleepers which will keep the steel straight and away from any rising dampness. Ideally this storage area should be under cover. Before use careful examination must be made for loose rust, mill scale, oil or grease, any of which can prevent bonding between the steel and concrete and result in structural failure. Loose rust and mill scale must be removed prior to use by wire brushing. However, where bars have a light coating of rust that is not loose or pitted, they may be used without treatment. Oil or grease on bars can be removed by using a proprietary degreasing agent and then thoroughly washing off with clean water.

Heavily pitted bars should be rejected as the resulting loss in cross-sectional area will cause loss of strength.

Testing

The testing of concrete and its constituent materials is carried out in order to determine a material's suitability for inclusion in a concrete mix and to ensure that the quality of the wet and hardened concrete is both acceptable and consistent.

When carrying out tests, whether on separate materials or on wet or hardened concrete, it is essential that these are performed using standard procedures. Many of these standard procedures are set out in relevant British Standards. The basic idea of any test is to examine the effect of one variable factor only. If the procedure adopted in carrying out a test is not consistent from one test to another then other variables will creep in, making any comparison of the results useless.

The tests covered in the following sections are

those which are most often carried out on site as a form of quality control. The methods used follow the standard procedures, although more detailed descriptions are contained in the relevant British Standards and it is considered essential that these are available on site for reference as required.

Cement testing

All cement is manufactured in accordance with a standard specification and the manufacturer will carry out tests to ensure that it complies with this specification. Therefore it is rarely necessary to carry out cement testing. However, when a cement is suspect it may be tested in accordance with the relevant British Standard. The properties most commonly tested are:

Consistence of a standard paste
Setting times
Soundness
Compressive strength.

This testing must be carried out in an extensively equipped laboratory, by personnel having considerable experience of this type of work. Thus only exceptionally will cement testing be carried out on site.

However, a practical test often carried out on site on suspect lumpy cement (caused by poor storage conditions) is to squeeze the lumps between two fingers. If they are powdered easily the cement is considered suitable for use. However, it is likely to have lost some of its strength and thus before use in structural work it should be proved suitable by a cube test.

Aggregate testing

Aggregates must be tested on site at regular intervals to ensure that their various characteristics and properties are in accordance with those specified for the work in hand. Whilst there is a large number of tests that may be applied to aggregates, the following are those most often carried out on site:

Cleanliness
Grading
Moisture content
Bulking
Density.

Samples for testing

When aggregate samples are taken for testing they should be representative of the whole load or stockpile. The sample should be collected using a scoop in at least ten small portions taken from different positions in the stockpile or at regular intervals during the unloading of a lorry. This main sample should then be thoroughly mixed and reduced to the amount required for testing by either quartering or the use of a riffle box.

Quartering (Figure 124) Form the aggregate into a flat topped heap, and divide into quarters using either a shovel or a quartering device. Return the two diagonally opposite corners to the stockpile and remix the other two together. Repeat the process until an appropriate sized sample is obtained.

Riffle box (Figure 125) This is designed to split a sample equally in half without any form of selection. The aggregate is tipped into the top of the container; the baffles divide it equally and discharge it into the two containers below. One of

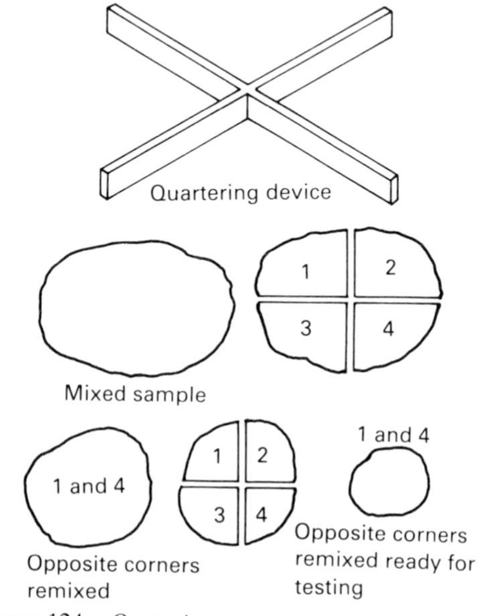

Figure 124 *Quatering*

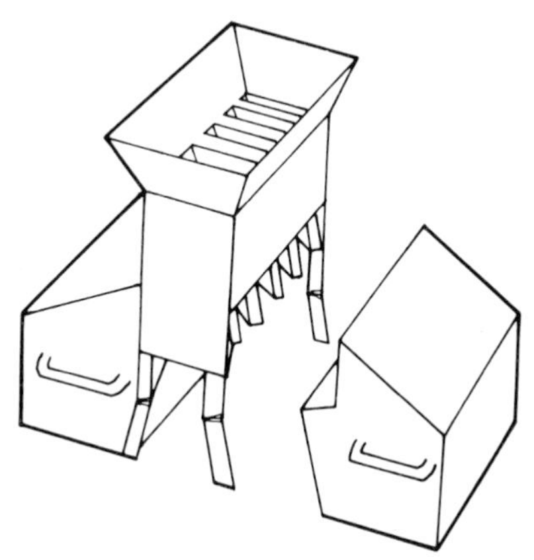

Figure 125 *Riffle box*

the containers is returned to the stockpile and the other is used to repeat the process until the required amount is obtained.

Cleanliness

A practical measure of aggregate cleanliness can be determined by rubbing a sample of the aggregate between the hands. If silt or clay remains on the hands it may be unacceptable for use and should be further tested.

Accurate tests for determining the proportions of clay, silt and dust in both fine and coarse aggregates are detailed in British Standards. However, these are only suitable for laboratory assessment. On site the amount of silt and other fine material present in sand can be fairly accurately determined using the field (on-site) settling test.

Field settling test

The following are the stages in a field settling test (Figure 126):

1 Prepare a 1 per cent solution of common salt in water (2 teaspoonfuls of salt to 1 litre of water).
2 Pour 50 ml of the solution into a glass measuring cylinder.
3 Add sand to the salt solution in the cylinder until the sand level itself reaches the 100 ml mark.
4 Top up the salt solution until it reaches the 150 ml mark.
5 Shake the cylinder vigorously, place on a level surface and gently tap it until the top of the sand is level.
6 Leave to settle for three hours.
7 The sand being heavier will settle first, leaving a layer of finer silt on top. Measure the height of the silt layer and the height of the sand (using the millilitre scale on the cylinder). The silt content is expressed as a percentage by volume of the sand content:

$$\text{silt content} = \frac{\text{height of silt}}{\text{height of sand}} \times 100 \quad \text{per cent}$$

Example

To determine the silt content of a sand sample after conducting a field settling test where the height of sand equalled 80 ml and the height of the silt equalled 6 ml.

$$\text{silt content} = \frac{6}{80} \times 100 = 7.5 \text{ per cent}$$

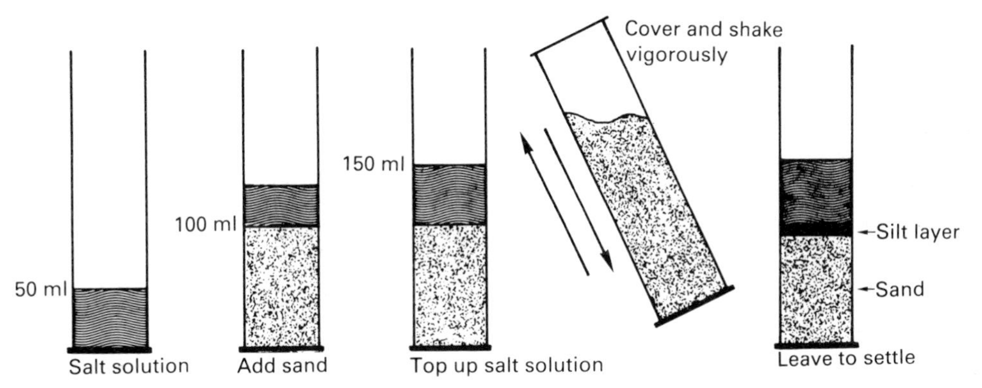

Figure 126 *Field settling test*

The aggregate is normally considered acceptable for use if the silt content does not exceed 8 per cent. For results in excess of this, samples should be referred to a testing laboratory for more detailed assessment.

Grading

To determine the grading of both fine and coarse aggregates it is necessary to carry out a sieve analysis. This involves shaking a sample through a series of sieves of standard mesh sizes. Those in general use are 75 mm, 63 mm, 37.5 mm, 20 mm, 14 mm, 10 mm and 5 mm for coarse aggregate, and 5 mm, 2.36 mm, 1.18 mm, 600 μm, 300 μm and 150 μm for fine aggregates. A lid and receiving pan is used in conjunction with the sieves in order to prevent accidental spillage and to collect the material passing through.

Sieve analysis

A sieve analysis is carried out as follows (Figure 127):

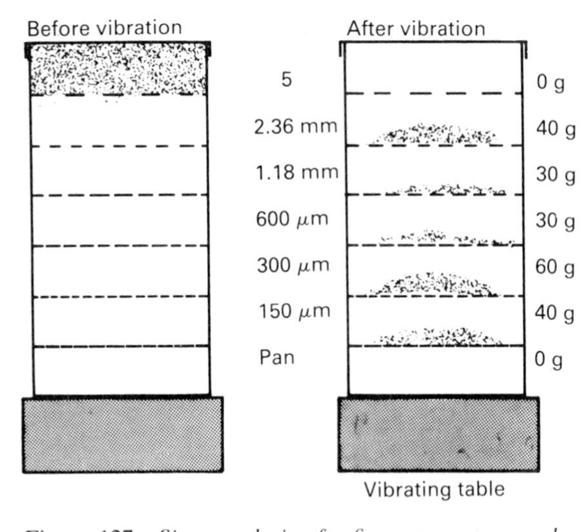

Figure 127 *Sieve analysis of a fine aggregate sample*

1 Weigh out a sample of air dried aggregate, say 2 kg for coarse aggregate or 200 g for fine aggregate.
2 Place the sample in the largest sieve which has the receiving pan fitted, fit lid and shake vigorously for at least two minutes.
3 Transfer the contents of the receiving pan into the next sieve size down (with a second receiving pan fitted). Assistance may be given to the large individual pieces of coarse aggregate by offering them by hand to the aperture to confirm whether or not they will pass through.
4 Repeat the procedure in 3 for the remaining sieve sizes in diminishing order.
5 Alternatively where a mechanical vibrator is available the complete nest of sieves can be assembled as one stack with a receiving pan on the bottom. The sample is placed in the top and the lid fitted, and the whole nest is then agitated for 10–15 minutes.

Complete a sieve analysis report form as illustrated in Figure 128. The percentage passing is calculated to the nearest whole number. On completion a check should be made that the total mass retained by the sieves and the final receiving pan is equal to the original mass of the sample. In practice a small difference is acceptable; where large differences occur the test should be repeated.

SIEVE ANALYSIS

MATERIAL	fine aggregate	REFERENCE	PS/CA 1
SOURCE	BBS Supplies	DATE	30–1–88
MASS	200 g	RESULT	Zone M

SIEVE SIZE	MASS RETAINED	MASS PASSING	PERCENTAGE PASSING
10.00 mm	—	—	—
5.00 mm	0	200	100
2.36 mm	40	160	80
1.18 mm	30	130	65
600 μm	30	100	50
300 μm	60	40	20
150 μm	40	0	0
PAN	0	0	0
TOTAL	200 g		

Figure 128 *Sieve analysis report*

The figures in the final percentage passing column can be compared with the permitted grading zones given previously for the aggregate being tested.

It is common to present these results visually in the form of a graph of sieve sizes horizontally against percentage passing vertically, as illustrated in Figure 129. The points plotted are joined up with straight lines to form what is termed a grading curve. The grading zone limits for the aggregate being considered can be added to this graph to determine whether or not the material is acceptable.

Example
To determine whether or not a sample of fine aggregate ordered as zone M (medium) does in fact fall within the permitted grading zone.

Carry out a sieve analysis and record the results using either of the methods illustrated in Figures 128 and 129. Compare the actual results with the grading zone limits. In this case the results fall within the zone, and the aggregate is therefore zone M as ordered.

Moisture content
Knowledge of the moisture content of aggregates is an important factor in achieving a concrete mix with a consistent workability and strength. The quantities of aggregate and water specified for a mix must be adjusted to compensate for the amounts of natural moisture being put into the concrete mixer with the aggregate. The three main methods that can be used to determine the moisture content are by drying a sample, by the use of a syphon can or by the use of a moisture meter.

Determination of moisture content by drying
This may be carried out by the oven drying method using the following procedure (Figure 130):

1 Weigh out a sample of the aggregate, say 2 kg for coarse aggregate or not less than 500 g for fine aggregate, and place in an airtight container and reweigh. This is recorded as the wet mass.
2 Remove the lid from the container. Place the container and its contents in a well ventilated oven controlled at 105°C for 16–24 hours.
3 Remove the container from the oven. Replace the lid and allow to cool for between ½ and 1 hour.
4 Reweigh the container and record this as the dry mass.
5 Calculate the moisture content as a percentage of dry mass using the formula:

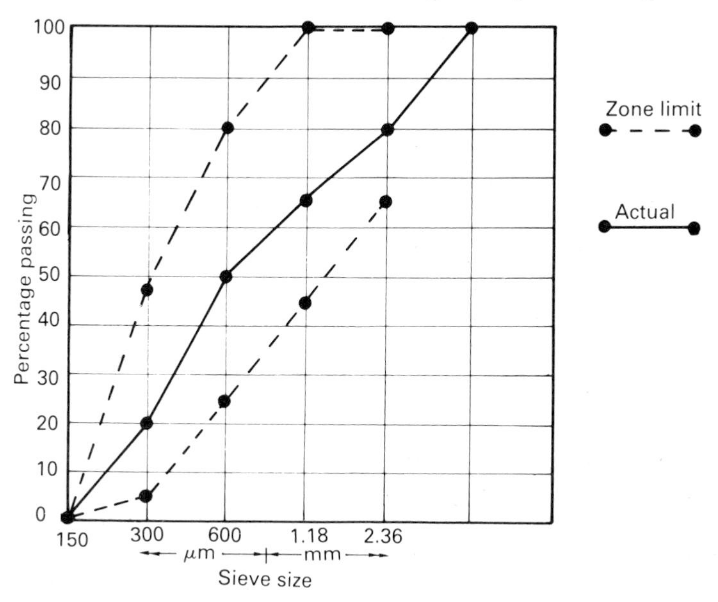

Figure 129 *Grading curve example*

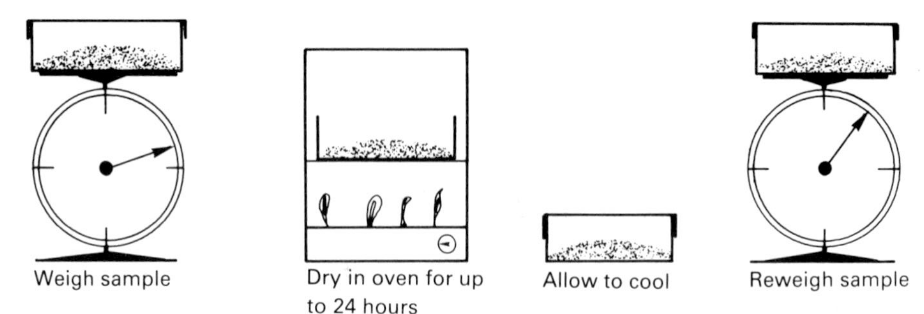

Figure 130 *Oven drying*

$$\text{moisture content} = \frac{\text{wet mass} - \text{dry mass}}{\text{dry mass}} \times 100 \text{ per cent}$$

Example
To determine the moisture content of a sample of aggregate having a wet mass of 2 kg and a dry mass of 1.85 kg.

$$\text{moisture content} = \frac{2 - 1.85}{1.85} \times 100 = 8 \text{ per cent}$$

Determination of free moisture content by drying
The previous method gives the total moisture content of the aggregate. The moisture content that is used when making adjustments to the amount of mixing water is known as the free moisture content. The aggregate is said to be in a saturated but surface dry condition. If the total moisture content were to be used for adjustments the aggregate pores would absorb some of the added water required for hydration, resulting in a dry, poor workability, weak concrete. Using the free moisture content the pores are already saturated and therefore will not absorb any of the added water.

This drying method can be carried out much more speedily (approximately 10–15 minutes) and is ideal for site use (Figure 131). The frying pan method (as it is commonly termed) is carried out using a similar sample of aggregate, and drying it in an open pan using the gentle heat of a gas stove or electric hairdryer. The sample should be stirred continuously whilst drying, until the glistening moisture on the aggregate particles appears to have evaporated. The percentage free moisture content is calculated from the formula:

$$\text{free moisture content} = \frac{\text{wet mass} - \text{surface dry mass}}{\text{surface dry mass}} \times 100 \text{ per cent}$$

Example
To determine the free moisture content of a sample of fine aggregates having a wet mass of 750 g and a surface dry mass of 714 g.

$$\text{free moisture content} = \frac{750 - 714}{714} \times 100 = 5 \text{ per cent}$$

Use this percentage to determine the amount of water to be added at the mixer.

Figure 131 *Frying pan drying*

Example
To determine the amount of added mixing water for a mix containing:

170 kg or litres of water
300 kg of cement
665 kg of fine aggregate
1200 kg of coarse aggregate

If the free moisture contents of the aggregates as determined by site tests are 5 per cent for fine and 2 per cent for coarse.

free moisture in fine aggregate
= mass of fine aggregate × free moisture %
= 665 × 0.05 = 33.25 kg or litres

free moisture in coarse aggregate
= mass of coarse aggregate × free moisture %
= 1200 × 0.02 = 24 kg or litres

water added at mixer
= total water content − free moisture content aggregates
= 170 − 33.25 − 24 = 112.75 kg or litres

Bulking
When a sample of aggregate is dry its particles are tightly fitted together and its volume is at a minimum. When the moisture content is increased the solid particles are separated by a film of liquid. The thickness of this film causes the aggregate to bulk or increase in volume. The actual increase in volume is dependent on the moisture content, which governs the film thickness of moisture around each particle, and also the particle size. In general bulking increases as the particle size decreases, owing to the greater surface area in a finely graded sample (see Figure 132).

This is also illustrated for fine aggregate by the graph in Figure 133. This shows that finely graded

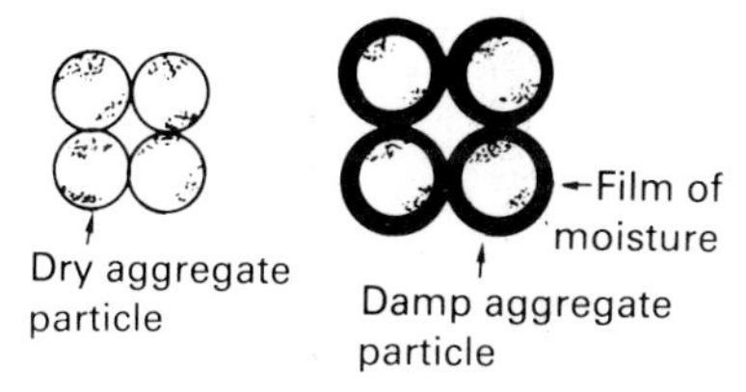

Figure 132 *Bulking of aggregate*

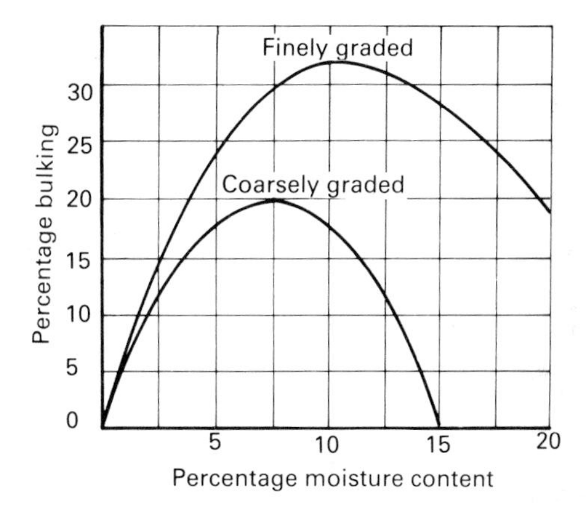

Figure 133 *Typical fine aggregate bulking curves*

aggregates bulk more than coarsely graded aggregates. When an aggregate is fully saturated, the films of water around each particle join up and the volume decreases again to that of a dry sample.

Bulking in coarse aggregate is largely inconsequential, owing to the comparatively small particle surface area in a sample and the fact that moisture can freely drain through the voids between adjacent particles and is not trapped in the material.

The amount of bulking in fine aggregates must be taken into account when batching a mix by volume, otherwise the mix will be deficient.

Bulking test for fine aggregate
This determines the percentage bulking of a sample of fine aggregate and is used to establish the amount of extra material needed in a mix (Figure 134).

1 Fill a measuring cylinder with damp, fine aggregate to about two-thirds of its height. Then tap the cylinder to consolidate and record the height of the aggregate (bulked volume), using the millilitre scale on the cylinder.
2 Half fill an identical measuring cylinder with water.
3 Pour the measured aggregate into the water cylinder; shake vigorously to expel air bubbles. Allow the aggregate to settle, then again

record the height of the aggregate (saturated volume).

4 Express the amount of bulking as a percentage from:

$$\text{bulking} = \frac{\text{bulked vol.} - \text{saturated vol.}}{\text{saturated volume}} \times 100 \text{ per cent}$$

Use this percentage figure to determine the amount of extra material required.

Example

To determine the amount of damp fine aggregate required for a concrete mix requiring 0.3 m^3 of dry fine aggregate, a sample in a measuring cylinder was 180 ml when bulked and 155 ml when saturated.

$$\text{bulking} = \frac{180 - 155}{155} \times 100 = 16 \text{ per cent}$$

Therefore the volume of damp fine aggregate must be increased by 16 per cent to allow for bulking:

$$\text{required damp volume} = \text{dry volume} + \% \text{ bulking}$$
$$= 0.3 + (0.3 \times 16/100)$$
$$= 0.348 \text{ m}^3$$

Note Percentage increases can be obtained by turning the percentage into a decimal. Place a decimal point behind the percentage and move two places forward. Place a 1 in front of the point (to include the original quantity) and multiply by the resulting figure. In the example, the required damp volume can thus be obtained as $0.3 \times 1.16 = 0.348$ m^3.

Density

The density of a material can be defined as its mass per unit volume, and is expressed in kg/m^3:

$$\text{density} = \frac{\text{mass}}{\text{volume}}$$

Bulk density

When measuring the volume of a sample of aggregate, the volume of air voids between the particles will also be taken into consideration. Therefore the density determined will only be the bulk or apparent density and not the absolute value.

$$\text{bulk density} = \frac{\text{mass}}{\text{bulk volume}}$$

The bulk density of an aggregate is determined from the mass of aggregate that just fills a container of known volume. The mass will vary depending on the amount of compaction used when the container is filled. Two values for bulk density may be determined:

Uncompacted bulk density or, as it is more commonly known, *loose* bulk density. This is found by filling a container to overflowing with a sample of aggregate, using a shovel or scoop and discharging from a height of not more than 50 mm above the rim. Strike off the surface material flush with the rim using a straight edge.

Compacted bulk density This is found in the same way as before except that the aggregate is tamped with a rod during filling to compact it. This method is not suitable for fine aggregates as the tramping tends to crush the particles, giving unreliable results.

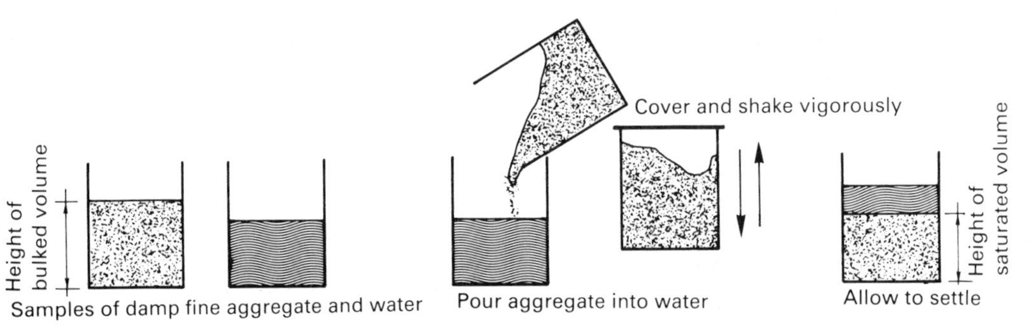

Figure 134 *Bulking of fine aggregate*

As the moisture content of the aggregate will affect the bulk density, either these tests should be carried out on oven dried samples or the moisture content to which the bulk density refers must be stated.

Example
To determine the bulk density of a sample of fine aggregate which filled a 5 litre container and had a mass of 8.525 kg.

$$\text{bulk density} = \frac{8.525}{0.005} = 1705 \text{ kg/m}^3$$

Note 1000 litres = 1 m^3; therefore 1 litre = 0.001 m^3.

Relative density
The relative density of any material is the relation of the material's density to the density of water. This is also known as the specific gravity.

The relative density of aggregates can be determined by carrying out tests detailed in British Standards, although for most purposes the standards values of 2.7 and 2.6 for crushed and uncrushed aggregates respectively are often assumed. This means that uncrushed aggregates are assumed to be 2.7 times heavier than water, and uncrushed aggregates are assumed to be 2.6 times heavier than water.

Concrete testing

Concrete must be tested periodically in both its fluid and hardened state to ensure that it complies with the specification. The most common tests carried out are for the workability of fluid concrete and the strength of hardened concrete.

Samples for testing
A sample of fluid concrete taken for testing must be representative of the whole batch. Wherever possible the sample is best taken in a number of increments as the concrete is being discharged from the mixer or ready mix truck (Figure 135).

For most tests four increments are taken periodically during the discharge, by passing a standard scoop across the stream of concrete flowing down the shute. These scoopfuls are collected in a barrow or bucket.

When carrying out a slump test using concrete from a ready mix truck it is permissible to take the sample in one increment after approximately 0.3 m^3 has been discharged. This enables the workability of the mix to be determined at an early stage, rather than when most of the concrete has been discharged.

Before testing all samples taken must be thoroughly remixed on a non-absorbent surface using a shovel, to ensure uniformity.

Workability
As stated previously the term 'workability' is used to describe the ease with which a concrete mix can be placed and compacted. Workability is not itself a measureable property. Instead tests for workability measure properties of concrete which vary with changes in the workability and thus give an indication of workability. These standard tests are:

Slump test
Compacting factor test
V-B consistiometer test.

Slump test
This is the most common test for workability and involves measuring the distance (slump) that a

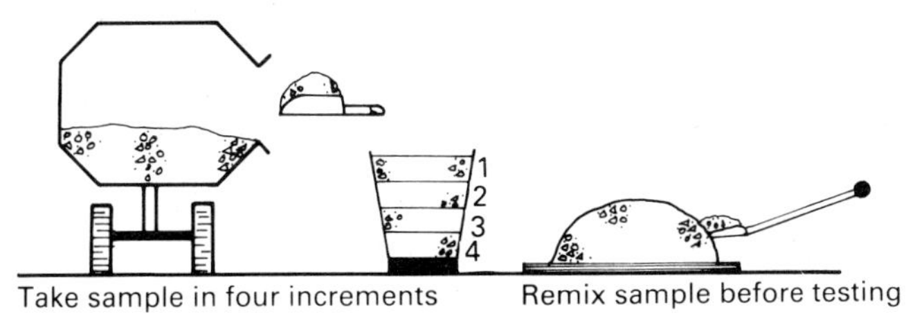

Figure 135 *Taking a sample for testing*

standard cone shaped mould of concrete collapses when its lateral support is removed (Figure 136). A concrete mix of low workability will have little or no slump, whereas a greater slump indicates a higher degree of workability.

The procedure is as follows:

1 Stand astride a clean dry standard slump cone with feet on the foot rests to hold the cone securely down.
2 Fill the cone with the sample mix in four separate layers, tamping each layer to its full depth with the tamping rod 25 times to ensure compaction.
3 Strike off excess concrete from the top of the cone using the tamping rod and clean the cone and base of any concrete spillage.
4 Slowly raise the cone vertically, avoiding any sideways movement.
5 Invert the cone and place it alongside the mould of concrete. Rest the tamping rod on top of the cone and measure the distance (slump) to the highest level of the partially collapsed concrete mould. This is expressed in millimetres to the nearest 5 millimetres.

In most situations the mould of concrete will collapse symmetrically when the cone is removed. However, there are three possible results as illustrated by Figure 137:

True slump A partially collapsed mould where the concrete merely subsides but still retains its approximate shape.
Shear slump Where one side of the mould shears off and collapses significantly more than the other.
Collapse slump Where the concrete collapses completely, totally losing its conical shape.

Only true slumps should be measured. If the mould shears or collapses the test should be

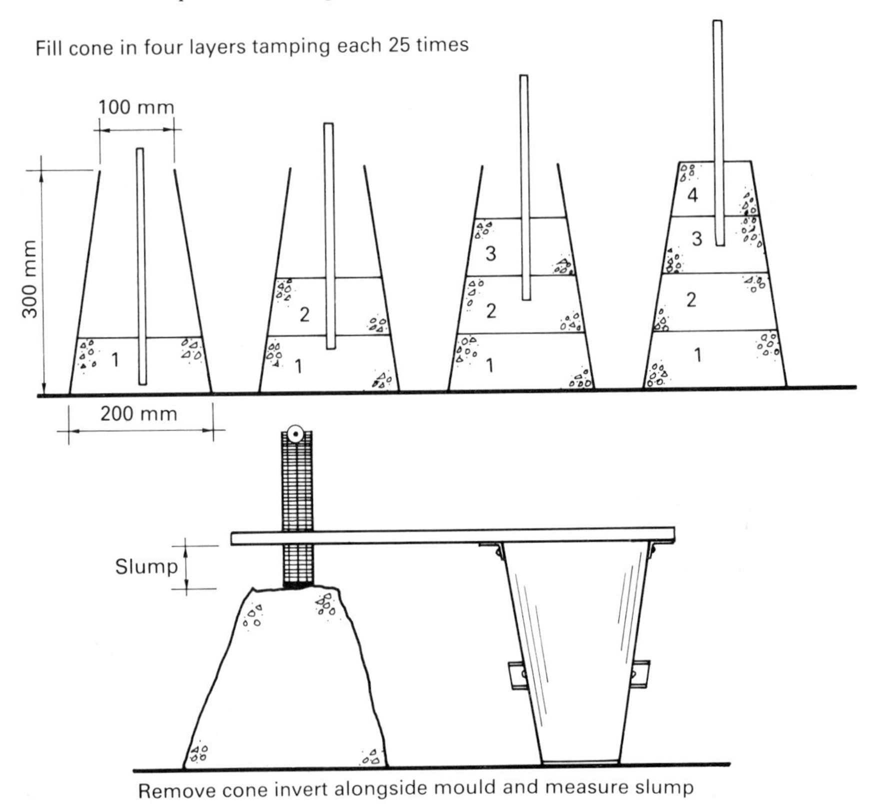

Figure 136 *Slump test*

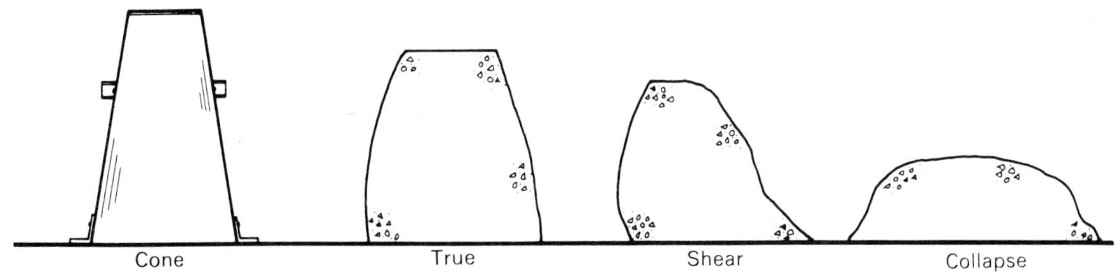

Figure 137 *Types of slump test*

repeated in an attempt to retain a true slump. If the second test shears or collapses again this fact should be recorded.

Compacting factor test

This test is particularly useful for assessing the workability of fairly dry mixes which have little or no slump. It measures the mass of a sample of concrete that has been partially compacted by a fixed amount of work and compares it with the mass of a fully compacted sample (Figure 138).

The procedure is as follows:

1 Ensure that the apparatus is clean, the trap doors are closed and the cylinder is covered.
2 Loosely fill the first hopper with the concrete mix.
3 Open the trap door at the bottom of the first hopper, allowing the concrete to drop into the second hopper (some compaction will take place).
4 Remove the cover from the cylinder. Open the trap door at the bottom of the second hopper, allowing the concrete to drop into the cylinder (still more compaction will take place).
5 Remove excess concrete from the top of the cylinder by slicing off using two steel floats.
6 Weigh the cylinder and contents and record as the mass of partially compacted concrete.
7 *Either* vibrate the partially compacted concrete in the cylinder until it is fully compacted, adding more concrete as required to top it up. Smooth the concrete at the top of cylinder with a steel float.
   Or empty the cylinder. Refill with the same mix in six separate layers, tamping each layer with a tamping rod at least 30 times to ensure complete compaction.

   Weigh the cylinder and its either vibrated or hand tamped contents and record as the mass of fully compacted concrete.
8 Determine the compacting factor from the following:

   compacting factor

$$= \frac{\text{mass of partially compacted concrete}}{\text{mass of fully compacted concrete}}$$

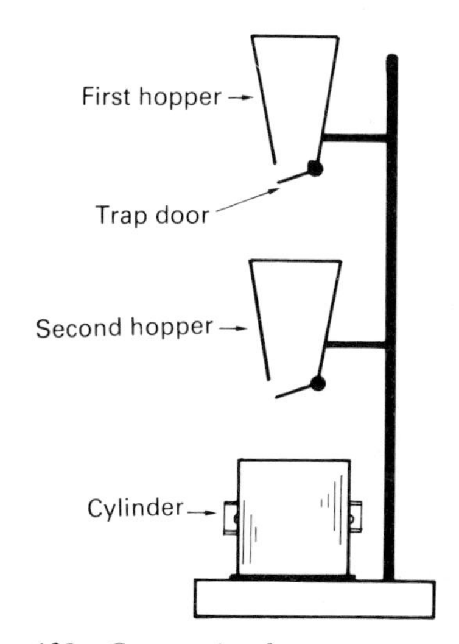

Figure 138 *Compacting factor apparatus*

The mass of the partially compacted concrete will always be less than that of the fully compacted concrete; therefore the resulting compacting factor (to two decimal places) will always be less than 1. Compacting factors will range between about 0.95 for highly workable mixes and 0.70 for low workability mixes.

Example
To determine the compacting factor of a sample having a partially compacted mass of 10 500 g and a fully compacted mass of 12 000 g.

$$\text{compacting factor} = \frac{10\ 500}{12\ 000} = 0.88$$

V-B consistometer test
This test more accurately measures the workability of fairly dry mixes having little or no slump than does either the slump or the compacting factor test. However, it is mainly restricted to the laboratory or sites where large volumes of high quality concrete are to be poured. It measures the time taken to compact a standard cone, full of concrete. The time taken in seconds (V-B degrees) is directly proportional to the amount of work carried out.

The procedure is as follows (Figure 139):

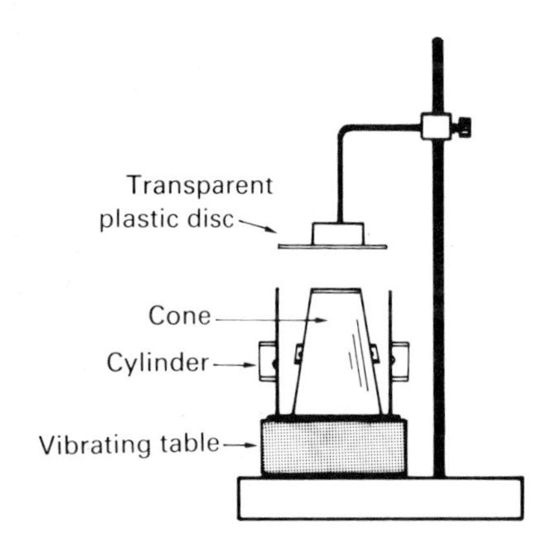

Figure 139 *V-B consistometers apparatus*

1 Fill and compact a standard cone (without feet) inside the cylindrical container which is attached to the vibrating table. The cone should be filled and compacted using the same method as that detailed for the slump test.
2 Withdraw the cone. If the slump is true it may be measured if required. Swing the transparent plastic disc into position over the cylinder and adjust until it just touches the concrete surface.
3 Switch on the vibrator and start the clock. The vibration will cause the cone shaped mound of concrete to remould itself to the shape of the container.
4 Stop the vibrator and record the time at the moment the transparent plastic disc becomes fully coated with concrete.

The recorded time to the nearest 0.5 seconds is termed V-B seconds or degrees. This will vary from 0 for high workability mixes to 25 or over for dry, low workability mixes.

Comparison of workability test results
There is no direct relationship between the different workability tests, since the tests are different and measure different properties. However, an approximate comparison can be made of slumps, compacting factors and V-B degrees to be expected from mixes of different workabilities. Table 21 is divided into the five concrete mix workabilities commonly referred to. For each it compares approximate workability test values and states typical uses.

Strength testing
Hardened concrete samples are strength tested to maintain a check on the quality of the concrete being produced. The most commonly carried out test is the cube test for compressive strength. Others include cylinder testing for indirect tensile strength and beam testing for flexural or bending strength.

Cube testing
This consists of making standard size concrete cubes; curing and storing them in a standard manner; and finally crushing them at 7 and 28 days, when they should achieve at least the minimum strength specified.

The procedure for making a cube is as follows (Figure 140):

1 Take a clean standard cast iron or steel cube mould, bolt the two halves together and secure to its base (150 mm cubes are normally

Table 21 *Approximate comparison of workability tests and typical end uses*

Workability	*Slump mm*	*Compacting factor*	*V-B degrees*	*Typical end use*
Extremely low	0	0.65–0.70	over 20	Very high strength concrete, normally only used in precasting factories, having extensive vibration facilities
Very low	0–25	0.70–0.80	7–20	High strength concrete sections and mass concrete. Must be compacted by vibration
Low	25–50	0·80–0.90	3–7	Hand compacted plain concrete sections and normal reinforced concrete sections compacted by vibration
Medium	50–100	0.90–0.95	1–3	Normal reinforced concrete sections compacted by hand and heavily reinforced concrete sections compacted by vibration
High	100–150	over 0.95	0–1	Normally only suitable for hand compacting heavily reinforced sections or other intricate sections which are particularly difficult to compact.

used for aggregates up to 40 mm) but 100 mm cubes may be used where the aggregate does not exceed 20 mm).

2 Lightly coat the internal surfaces of the cube mould with a film of release agent.
3 Make a representative sample of the mix, using the method outlined previously.
4 Fill the cube mould with the sample mix in 50 mm layers, tamping each layer with the tamping bar at least 35 times for 150 mm cubes and 25 times for 100 mm ones. Alternatively the sample may be compacted by vibration.
5 Smooth the top of the cube with a steel float and clean any concrete spillage from the outside of the mould.
6 Cover the mould with damp sacking and a polythene sheet. Store for 16–24 hours at a temperature of 20°C.
7 Demould the cubes. Mark them with a wax crayon for later identification. Submerge them in a water tank maintained at a temperature of 20°C until required for testing.

Crushing a cube

The test is normally carried out off site by a specialist concrete testing laboratory. It requires a purpose-made machine conforming to British Standards that is capable of applying a compressive load at a constant rate of increase.

The procedure is as follows (Figure 141):

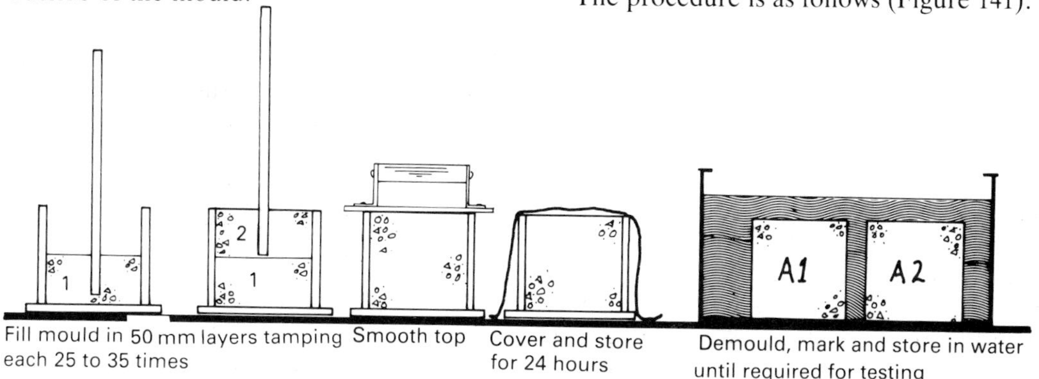

Figure 140 *Cube test*

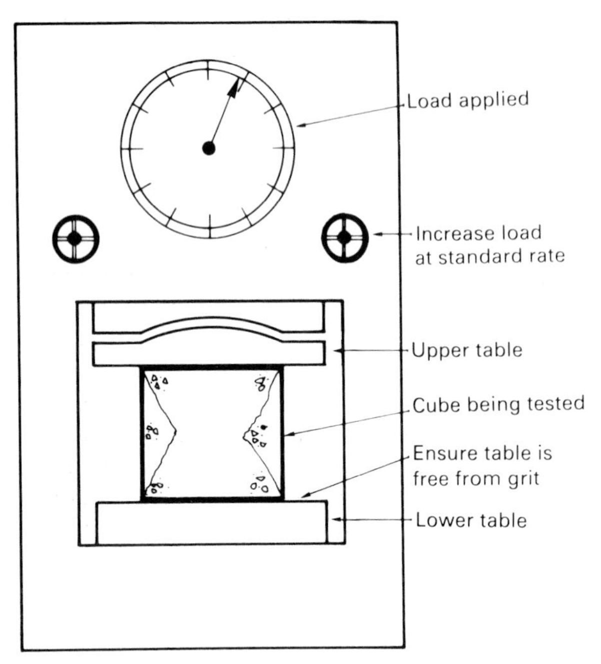

Figure 141 *Cube crushing*

1 Remove the cube from water just prior to testing.
2 Ensure that the machine and cube to be tested are free from any loose grit.
3 Place the cube centrally in the testing machine so that the floated surface which was on top when the cube was cast is now vertical.
4 Apply and increase the load at a standard increasing rate.
5 Record the maximum load applied before failure and note the type of failure.
6 Determine to the nearest 0.5 N/mm^2 the crushing strength of the cube from the following:

$$\text{crushing strength, } N/mm^2 = \frac{\text{load at failure in N}}{\text{area in } mm^2 \text{ of one cube face}}$$

It is normal practice to prepare and test two cubes from a sample and to use the average crushing strength of the two tests as the actual result.

Example
To determine the crushing strength of a 150 mm concrete cube that failed at 725 KN.

$$\text{crushing strength} = \frac{725 \times 1000}{150 \times 150} = 32 \text{ N/mm}^2$$

The 1000 in the numerator converts kilonewtons to newtons.

Inaccurate crushing strength results can occur if the cube has been poorly made, incorrectly cured, or misplaced in the testing machine.

The types of cube failure are shown in Figure 142. Satisfactory failures are those where all four exposed faces are cracked approximately equally, with little or no damage to the upper and lower faces in contact with the testing machine. Any other results are considered unsatisfactory.

Testing cast concrete
The strength tests which are carried out on samples of concrete are intended to give an indication of the strength of concrete used on site in a particular location. However, the test and actual strengths may vary due to differences in the amounts of compaction, the methods of

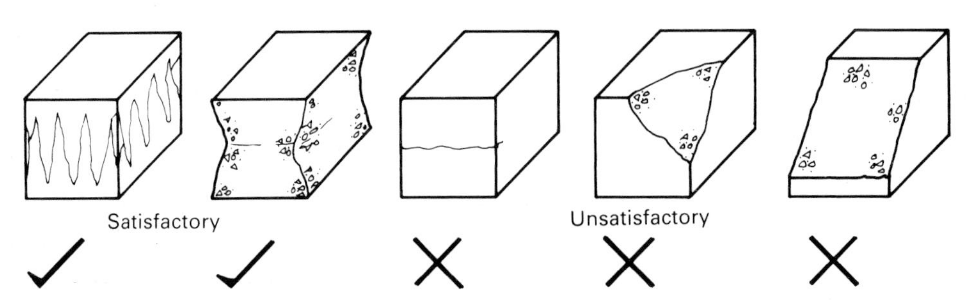

Figure 142 *Types of cube failure*

curing, the cube and structure dimensions, and the type of load.

Where strength tests on samples prove unsatisfactory it may be necessary to carry out testing of the actual cast concrete in order to determine its acceptability or otherwise. These tests include: cutting a cylindrical core from the cast concrete and submitting it to an equivalent cube test; a rebound hammer test, which measures the surface hardness of the concrete and gives an approximate indication of its strength; and full scale loading of the structure to determine the extent of deflection.

In addition, cast reinforced concrete may be checked to determine whether or not the minimum amount of concrete cover to the steel reinforcement complies with the specification. This is carried out using an electromagnetic cover meter. These are able to measure cover up to 100 mm and also to indicate the direction of the steel reinforcement.

Steel testing

No site or laboratory testing is required for reinforcement materials as they will have been manufactured in accordance with the relevant British Standards. The manufacturer will have carried out tests to ensure that the standards are met. However, on a practical level a visual inspection of reinforcement should be made to ensure that it is free from any contamination that could affect its subsequent bonding to the concrete or result in a loss of strength, e.g. loose rust, mill scale, oil, grease, mud and pitting.

Self-assessment questions

Question — *Your answer*

1 The purpose of a slump test is to determine:
(a) Moisture content
(b) Workability
(c) Strength
(d) Bulk density.

a [ ]	b [ ]	c [ ]	d [ ]

2 Which *one* of the following factors would cause a reinforcing bar to be rejected?
(a) Loose rust
(b) Pitting
(c) Mud
(d) Grease.

a [ ]	b [ ]	c [ ]	d [ ]

3 A batch of steel reinforcement is lightly coated in rust on delivery. It is recommended to:
(a) Reject the delivery
(b) Treat with a rust inhibitor
(c) Brush it off
(d) Use as it is.

a [ ]	b [ ]	c [ ]	d [ ]

4 The effect of finely grinding a cement is to:
(a) Increase its resistance to frost
(b) Improve workability
(c) Reduce early heat
(d) Increase strength development.

a	b	c	d
[ ]	[ ]	[ ]	[ ]

5 Describe *two* alternative procedures for taking a sample of concrete from a ready mix truck.

6 (a) Describe briefly the term *aggregate grading*.
(b) Distinguish between fine and coarse aggregate.

7 (a) Briefly describe the procedure for carrying out the compacting factor test on fresh concrete.
(b) State the purpose of this test.

Chapter 6

Concrete production and handling

Concrete can be produced on site from its constituent raw materials (cement, aggregates, water and any admixtures). Alternatively it may be produced at one of the ready mix plants located throughout the country and transported to the site in mixer trucks.

Storage of materials

Aggregates

Aggregates (Figure 143) are normally supplied in bulk, each size being stored separately adjacent to the mixer. Ideally these stockpiles should be on a hard concrete base laid to falls, and separated into bays by division walls. Typical division walls are H section reinforced concrete or steel vertical stanchions with timber or concrete planks located between them. Prior to delivery to site the aggregates will have had little protection from rain or snow; thus they will probably contain an excess of water. This excess water, which should be allowed to drain down through the stockpile overnight after delivery and prior to its being used, has the useful effect of removing silt or clay from the stockpile. The bottom 250 mm to 300 mm of a stockpile tends to remain saturated, and in any case will be heavily contaminated and thus should not be used for concrete work.

On delivery, prior to tipping, aggregates must be inspected to ensure that they are of the correct type and size and that they are clean. The type and size can be determined by eye. Cleanliness can be judged by rubbing a sample of the aggregate between the hands. If silt or clay remains on them, further tests should be carried out to determine whether or not to accept the material. Any doubts with regard to the size, type and cleanliness of a delivery should be referred to the site engineer for a decision.

Stockpiles should be sited away from trees to prevent leaf contamination. They should also be kept free from general site and canteen rubbish. Tarpaulins or plastic covers are useful for protecting stockpiles from rubbish and rainwater. In severe winter conditions the use of insulating blankets is to be recommended as these provide some protection from snow and frost.

Cement

Cement is supplied by its manufacturer either in bulk or in bags (see Figure 144).

Bulk cement

This is delivered by bulk tanker lorries and transferred using compressed air into storage silos. Various silos are available with capacities of between 10 tonnes and 50 tonnes. These should be located adjacent to the mixing plant in an elevated position to allow discharge into the mixer by gravity. As cement does not flow readily, aerators are often used in silos to aid the cement flow.

Bagged cement

This should be stored in a ventilated waterproof shed on a sound dry floor. The bags should be stacked clear of the walls and no more than eight to ten bags high. This prevents bags becoming damp through any defect in the outside wall,

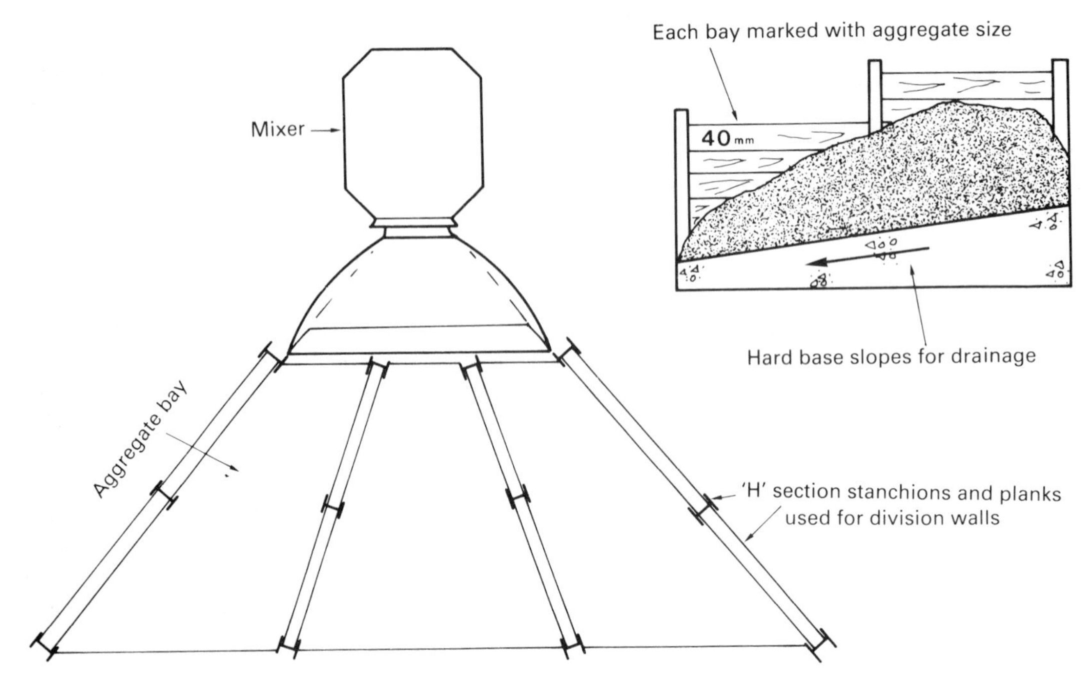

Figure 143 *Aggregate storage*

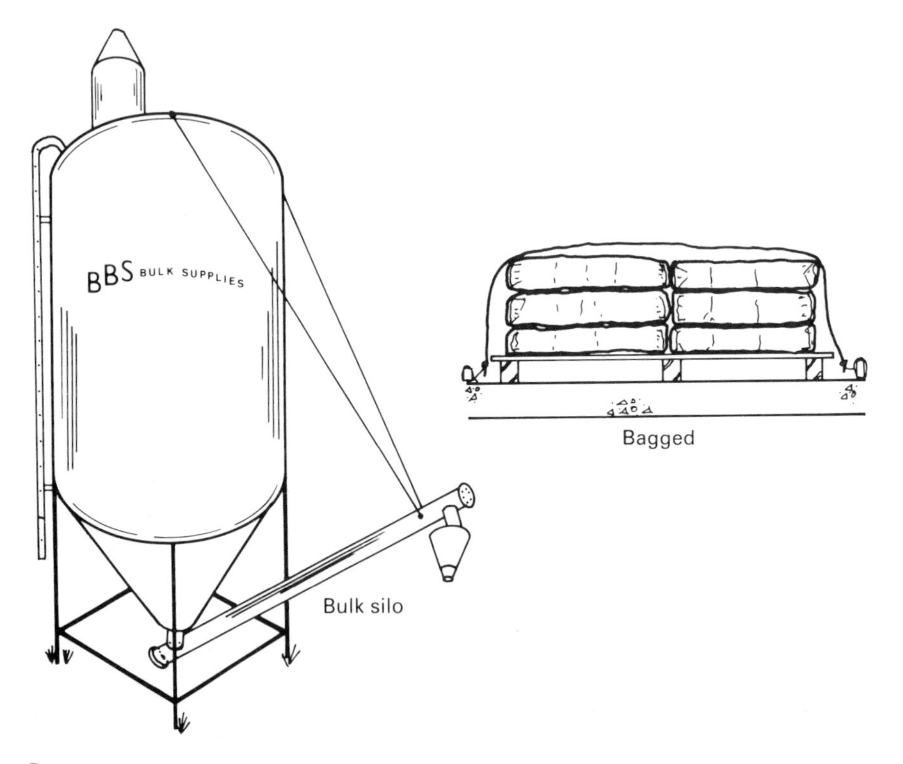

Figure 144 *Cement*

causing premature setting in the bag, and also compaction 'warehouse setting' of the lower bags through excessive weight from those above. Bags should be used in the same order as they were delivered. This is known as 'first in, first out'. Its purpose is to minimize the storage time, thus preventing any bags from becoming stale or 'air setting'. Bagged cement should remain in good condition for 4 to 6 weeks. However, any torn, damp or hardened bags, or cement that has become lumpy, should not be used for structural purposes. A practical test to determine the suitability of lumpy cement is to squeeze the lumps between two fingers. If they are powdered easily the cement is still usable but will have probably lost some of its strength. Thus still avoid structural use unless a cube test determines that a suitable strength can be achieved.

Where fairly small amounts of cement are being stored and a storage shed is not available, the bags may have to be stored in the open. They should be stacked six to eight bags high on timber pallets raised off the ground. The stack must be well covered with either tarpaulins or plastic sheets which are weighted down on top and at ground level.

Water

This may be supplied direct to the concrete mixer from the mains, via a preset automatic measuring water valve. Alternatively the water main can supply an elevated water storage cistern which in turn feeds a water measuring tank coupled to the mixer. Insulation of pipework and cisterns/tanks as well as a means of heating water should be considered for winter working.

Admixtures

These should be stored in the containers as supplied by their manufacturer's. Many admixtures are susceptible to extremes of temperature, exposure to which can change their properties in use. Thus they should be stored in conditions recommended in their manufacturer's instructions.

Batching

Accurate proportioning of the constituent materials is essential in order to produce concrete with specified properties. The process of measuring material quantities is known as batching. This may be carried out by volume or weight.

Batching by volume

This is suitable only for small quantities of concrete, mixed by hand or a small manually loaded mixer. The actual amount of cement and aggregate in any given volume will vary widely depending on its degree of compaction. However, if carried out with care using gauge boxes, buckets or other suitable containers, any inaccuracies and resulting variations in both workability and strength should be kept to a minimum (Figure 145). Wheelbarrows or shovelfuls must not be employed as a measure for volume batching as their contents will vary widely, making the production of a consistent mix impossible.

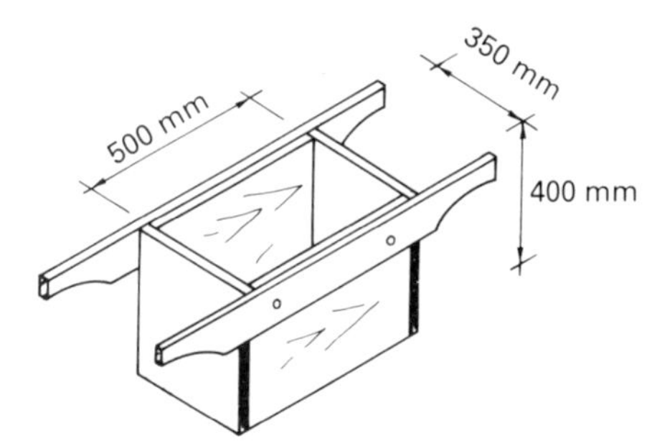

Figure 145 *Batching by volume*

Batching by weight

This is the most accurate method of batching and is almost universally used by most mechanical mixing plants. Weight batching of aggregates can be achieved by using either a mixer that incorporates a weighing device in its loading hopper, or a separate weight batching hopper that discharges into the mixer drum. Both types of batcher are loaded by means of a drag feeder shovel. This is a manually guided shovel, the forward movement of which is power assisted via steel wires attached to the mixer. Alternatively in large mixing plants aggregates may be stored in overhead hoppers which feed precise amounts of material via conveyor belts to the mixer. (See Figure 146.)

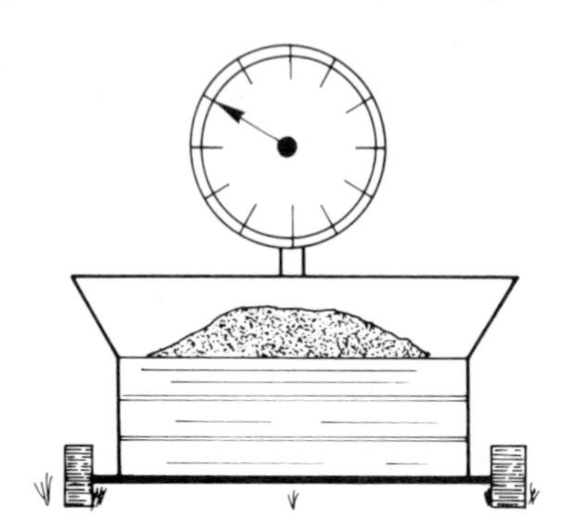

Figure 146 *Batching by weight*

Bagged cement is easily weight batched, each bag being 50 kg. Bulk cement silos normally incorporate weighing devices which deliver the required preset quantity of material. In addition both bagged and bulk cement may be weighed in a mixer's feed hopper.

Mixing

The batched constituent materials must be thoroughly mixed together to ensure that each is dispersed evenly throughout the mix. The aim is a homogeneous even-coloured material in which each particle of aggregate is evenly coated with the cement paste. There are a wide variety of concrete mixers in current daily use. However, these may all be classified into two types – continuous or batch.

Continuous mixers

These are capable of delivering a continuous supply of concrete without any stops for loading or discharging. Their use is in the main limited to large scale civil engineering projects, such as motorway construction and harbour work, where large quantities of concrete are required.

Batch mixers

These produce concrete in intermittent batches. Each batch of fully mixed concrete must be completely discharged before the following batch of constituent materials can be loaded. As the majority of building and civil engineering projects do not require the continuous supply of concrete for long periods, the batch type mixer is more often used in these situations.

There are a number of types of batch mixer which are classified in BS 1305 for batch type concrete mixers. These are designated by a number which represents the quantity of completely mixed concrete that the mixer can efficiently produce (not the bulk volume of materials loaded into the mixer), and a reference letter indicating the type of mixer, e.g.

T tilting drum
NT non-tilting drum
R reversing drum
P forced action (pan or paddle mixer)

For small mixers the number will come before the reference letter and will specify the output in litres. On larger mixers (over 1000 litres or 1 cubic metre) the output will be specified in cubic metres and be shown after the reference letter. For example:

100T = a tilting drum mixer with an output of 100 litres
R2.0 = a reversing drum mixer with an output capacity of 2 cubic metres

The normal range of sizes for these types of mixer is:

100T to T7.5
200NT to NT3.0
200R to R7.0
200P to P3.0

Tilting drum mixers

These have one opening in their cylindrical drum through which both the loading of the material and discharge of the concrete take place (Figure 147). Loading takes place from one side of the mixer with the drum opening horizontal; mixing is with the opening tilted upwards from the horizontal to avoid spillage. Discharge is from the other side of the mixer with the opening pointing downwards. The mixing action of a tilting drum mixer is by a pair of blades which lift the materials to the top of the drum as it revolves. When the materials reach the top they fall back to the bottom of the drum (encouraging mixing) to repeat the process. This free-falling action, which is common to all drum type mixers, has resulted in them being termed free-fall mixers.

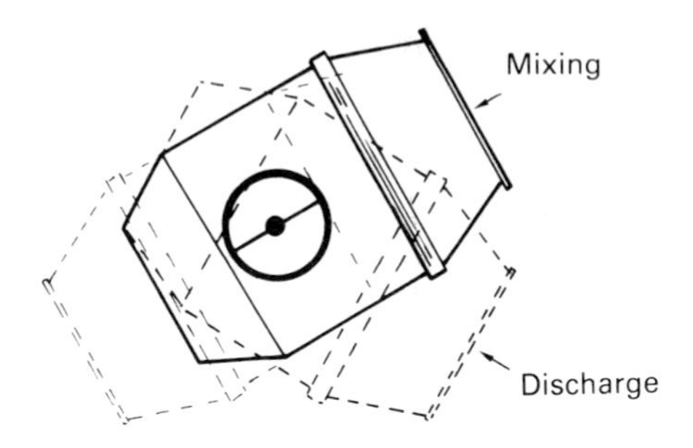

Figure 147 *Tilting drum mixer*

Non-tilting drum mixers
These have two openings pointing horizontally, one at either end of the drum (Figure 148). The material is loaded at one end and discharged from the other. The blades have the same action as the tilting in that they lift the material to the top of the drum and allow it to free fall. Discharge of the concrete is achieved by swinging a chute into the drum so that it catches the mixed concrete as it falls from the top of the drum.

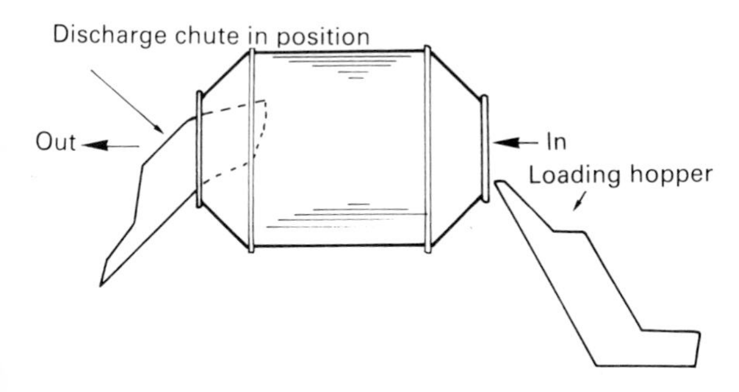

Figure 148 *Non-tilting drum mixer*

Reversing drum mixers
These are similar to the non-tilting type except for their method of discharge (Figure 149). Mixing is by a set of angled blades which do not only lift the mixture in the drum but also screw it forward along the drum during the mixing cycle. A second set of blades is so positioned in the drum that when the direction of rotation is reversed the blades guide the concrete out of the rimmed discharge opening.

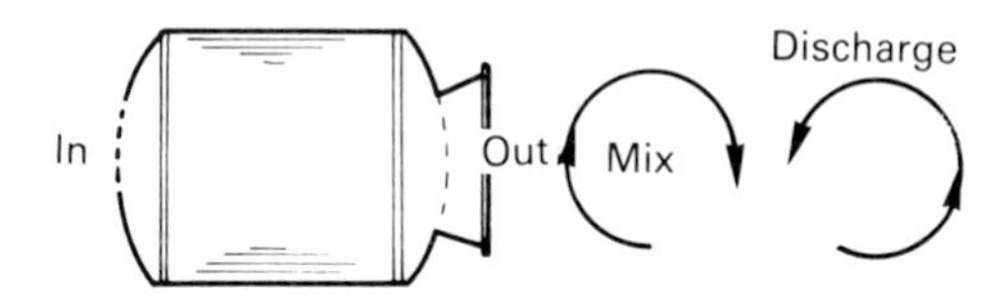

Figure 149 *Reversing drum mixer*

Forced action mixers
These are commonly known as either pan or paddle mixers. Their mixing action is forced mechanically throughout the mixing cycle, resulting in a quicker and more efficient method of mixing concrete (Figure 150):

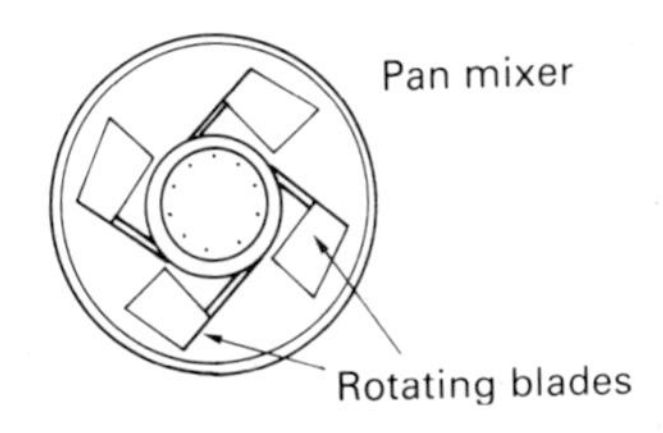

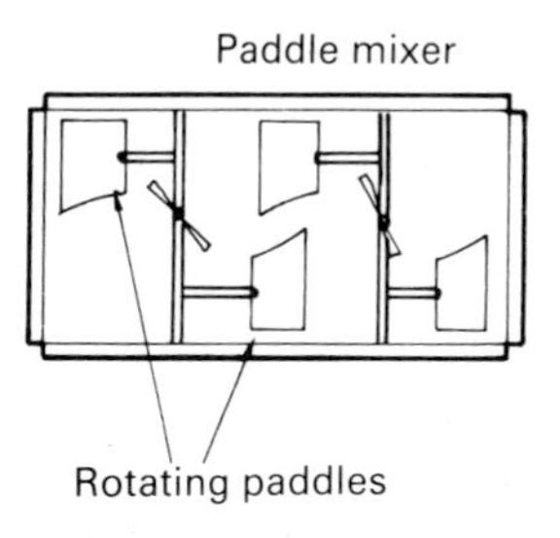

Figure 150 *Forced action mixers*

Pan type mixers consist of a circular horizontal pan which is fitted with internal blades. Loading is via the open pan top. Mixing is normally achieved by rotating the blades within the stationary pan. Alternatively both the pan and blades can be made to rotate on some mixers.

Discharge is via a door in the pan bottom which is kept closed during the mixing cycle.
Paddle type mixers consist of a rectangular trough across which are mounted horizontal counter-rotating paddles to achieve the mixing action. Loading and discharge is the same as the pan type mixer.

Ready mixed concrete

The majority of *in situ* cast concrete used in the UK (upwards of 80 per cent) is produced by ready mixed concrete plants located throughout the country. In general it is only very small and very large projects that use concrete mixed on site. Around two-thirds of ready mixed concrete is produced under the British Ready Mixed Concrete Association (BRMCA) authorization scheme. This scheme, which is regulated by the Association, sets minimum standards for the production of ready mixed concrete.

Ready mixed concrete can be produced in two different types of plant:

Dry batching plants Aggregates and cement are loaded into a truck mixer at the plant. Water is added either at the plant or on site and the concrete is mixed in the truck. The majority of ready mixed concrete comes from this type of plant.
Wet batching plants Aggregates, cement and water are mixed in a central mixing plant before being transferred to the truck mixer. The truck is used only as transport and an agitator for the concrete.

The truck mixer (Figure 151) consists of an inclined reversing drum that can be rotated at various speeds on a truck chassis. Both loading and discharge take place through a single hole located at the rear of the vehicle. An extendable chute that can be swung to point in different directions is provided to enable the concrete to be discharged in the required position. Truck mixers are fitted with a water tank, but this is only intended for washing out the mixer after discharge and not normally for increasing the water content of the mix. Extra water could be carried for this purpose, but will reduce the concrete carrying capacity of the vehicle.

Truck mixers of various capacities are available. Like their stationary counterparts they are designated by the letters TM followed by the output capacity in cubic metres. TM 2.0 to TM 10.0 vehicles are available. The actual amount of concrete that can be legally carried in a truck

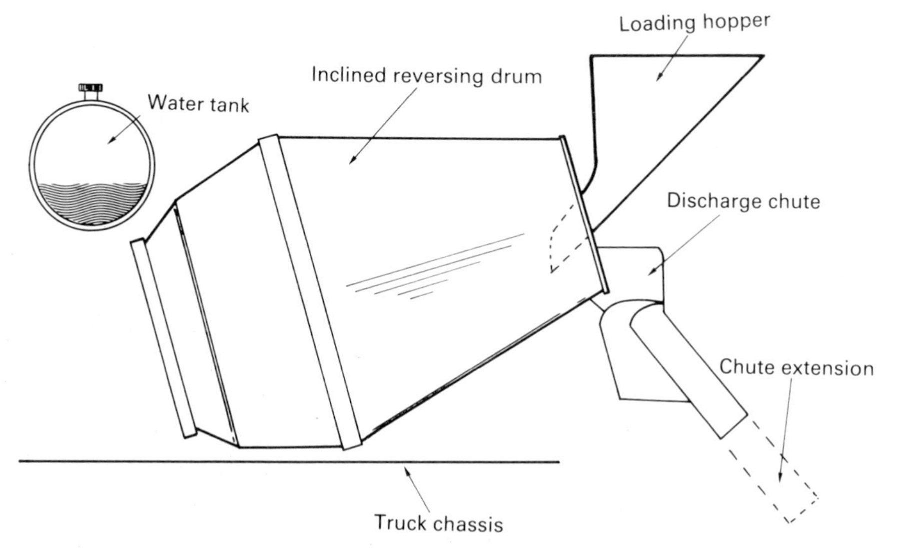

Figure 151 *Truck mixer*

mixer is governed under legislation concerning the permitted gross vehicle weights. The 24 tonne three-axle mixer truck which is used by the majority of the ready mix operators can legally mix and transport a maximum of 6 cubic metres of concrete.

Where ready mixed concrete is to be used on site, provision for access of the truck mixer must be made. Well made access roads are required, capable of supporting a 24 tonne fully loaded truck right up to the intended discharge point. Its size and manoeuvrability must also be considered. Typical dimensions are 8 m overall length, 2.4 m overall width, 3.45 m overall unladen height, and a minimum turning circle of about 15 m.

Failure to ensure adequate access prior to delivery can result in costly delays. Most ready mix operators allow 30 minutes from arrival on site to complete discharge. Any time spent on site in excess of this – normally due to a slow rate of placing, problems with reinforcement of formwork, or insufficient access – will be subject to an extra charge at a previously agreed rate.

The use of ready mixed concrete may be considered as a back-up on sites which normally mix their own concrete, e.g. where a failure has occurred with the on-site plant, or where the requirements of, say a massive concrete foundation exceed the site's output capacity.

Plant arrangements

The decisions on whether to use on-site produced concrete or ready mixed and, where on-site methods are selected, on the choice of plant to use, will depend on a number of points, including:

1 The total volume of concrete required
2 The amount of time available in the contract programme for concreting operations
3 The nature of the project and expected concrete strength and finish
4 The space available on site for concrete production plant and storage of materials
5 Whether concrete production plant is already owned or it has to be purchased or hired
6 The relative cost of on-site mixed concrete (taking into consideration costs of plant, labour and materials) compared with that for ready mixed concrete.

If on-site production is decided upon, the actual choice of mixer size to use is dependent on the mixing cycle time (time taken to batch and load materials, mix and discharge concrete, plus any delays) and the maximum hourly, daily or weekly concrete requirements. The actual mixing time of most drums is only between 1.5 and 2.5 minutes, but when the batching, loading and discharge times are added the shortest possible mixing cycle is between 4 and 5 minutes, with 6, 7 or 8 minutes being quite common. This gives between 7.5 and 15 complete mixing cycles per hour.

The minimum size of mixer can be calculated using the following formula:

$$\text{minimum output capacity of mixer} = \frac{\text{maximum concrete requirements, m}^3\text{ per hour}}{\text{mixing cycles per hour}}$$

The resulting figure can then be rounded up to the nearest available size of mixer.

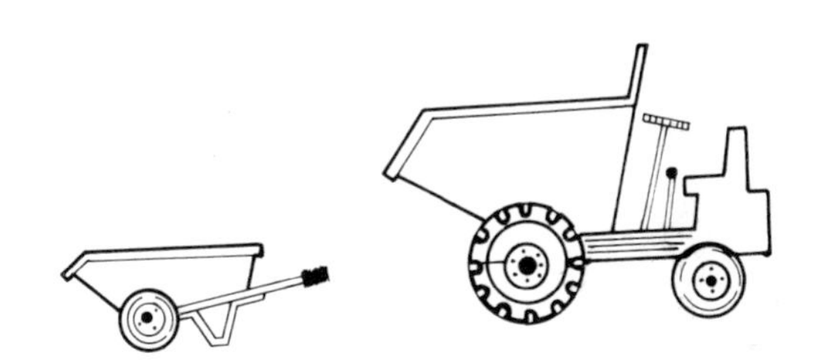

Figure 152 *Horizontal movement (pram barrow and dumper truck)*

Figure 153 *Horizontal movement (truck mounted transporter)*

Example
To calculate the minimum size of mixer where a maximum of 4.8 cubic metres of concrete per hour are required and the estimated number of mixing cycles per hour is 10.

minimum output capacity of mixer

$$= \frac{\text{maximum concrete requirements, m}^3\text{/hour}}{\text{mixing cycles per hour}}$$

$$= \frac{4.8}{10} = 0.480 \text{ m}^3 \text{ or } 480 \text{ litres}$$

This is rounded up to a suitable size, say 500 litres output capacity.

Care of mixers

After use each day, mixers must be thoroughly cleaned and maintained. This involves hosing off the outside of the plant; washing out the drum and cleaning off the blades to prevent a build-up of hardened concrete; cleaning out batching hoppers and checking their weighing accuracy; and periodically carrying out lubrication, preventive maintenance and inspection procedures in accordance with the manufacturer's handbook. Following this routine ensures that the plant is working at peak efficiency; reduces the likelihood of producing poor quality concrete; and minimizes the possibility of plant breakdowns.

Transporting and placing

Virtually all concrete will have to be transported on site from where it was mixed to its final location. This movement may be either horizontally or vertical or a combination of both. Owing to concrete's setting and hardening properties it should be transported from the mixer to the formwork in the shortest possible time. Additives to retard the setting of the concrete may be specified for the mix where delays are anticipated, thus extending the period available for transporting and placing. Long distance wheeled transport over bumpy sites causes segregation in the mix. This is the separation of the large and small particles in the mix, resulting in a weak non-uniform concrete with a soft surface layer. The very wet or very dry mixes are the most prone to segregation. Segregation may also occur in a mix that is permitted to fall freely any great distance. In general a drop of 2 to 3 m is the maximum unless some form of chute or pipework is used. The actual container in which the concrete mix is transported must be water tight so as to prevent the loss of fine material and grout, resulting in honeycombing.

Where concrete is being transported in open topped lorries or dumper trucks it is essential that they are covered with tarpaulins or similar at all times, irrespective of the weather conditions prevailing. This is because, if they are left uncovered, rain can increase the water content of the mix; sunshine can evaporate surface moisture; freezing conditions can cause rapid heat loss; and lastly the air current produced as the vehicle travels foward can itself cause evaporation.

Horizontal movement

Various forms of horizontal movement are shown in Figures 152 and 153.

Barrows and prams

These are the simplest type of transport, but are suitable only for small quantities over short distances. The larger two-wheeled prams and the petrol or diesel engine mechanized prams are more efficient in terms of volume of concrete moved per hour than are the single-wheeled barrows. The advantage of barrows and prams is their flexibility of movement. They are capable of entering small openings, being transported vertically in hoists and being wheeled along scaffolding.

Barrow runs over rough sites should be formed using scaffold boards to ensure a smooth route, thus reducing jolting and vibration and minimizing the risk of segregation. Single-wheeled barrows only require a run one board wide, whereas prams will require two, one for each wheel.

Dumpers and trucks

The most common method of transporting concrete on the average site is by means of a dumper.

These range in skip capacity from 250 litres to 2 cubic metres, and also in the method of skip discharge.

Some have a forward tipping discharge action that releases its load in one unstoppable movement when its restraining lever is released. An alternative to this is the hydraulic tipping skip that allow the operator to closely control the rate of discharge, which may be (depending on the model) directly forward, sideways, or through a turntable action that permits tipping at any angle. Some of these skips are fitted with a lifting mechanism which allows high level discharge directly into raïsed formwork up to about 1.5 m above ground level. Removable skip dumpers are available for use where batches of concrete require horizontal ground transportation followed by vertical crane handling.

Rough terrain fork lift trucks
These are a fairly recent development on building sites. When used for concrete transportation and placing they can be equipped with a detachable skip having a capacity of between 250 and 500 litres. In addition to horizontal movements, these trucks are also able to raise their skip vertically and discharge concrete up to a height of around 7 metres above ground level. This makes them extremely useful for placing concrete directly into walls and column formwork.

Larger amounts of concrete (in excess of 2 cubic metres) for mass pours, and road contracts etc. can be transported in truck mounted transporters, fitted with remixing blades at the point of discharge and a hydraulically operated tipping mechanism; in flat open top lorries with side or rear tipping action; or finally in the ready mix truck previously mentioned, which is able to mix and/or agitate its load of concrete.

Vertical movement
The available forms of vertical movement are shown in Figures 154 and 155.

Crane skips
These are available in a range of sizes. The largest in common use for building work are those having a capacity of up to 1 cubic metre. However, larger skips and buckets are available for mass pours and civil engineering work. The two basic types of skip are the vertical and the roll-over.

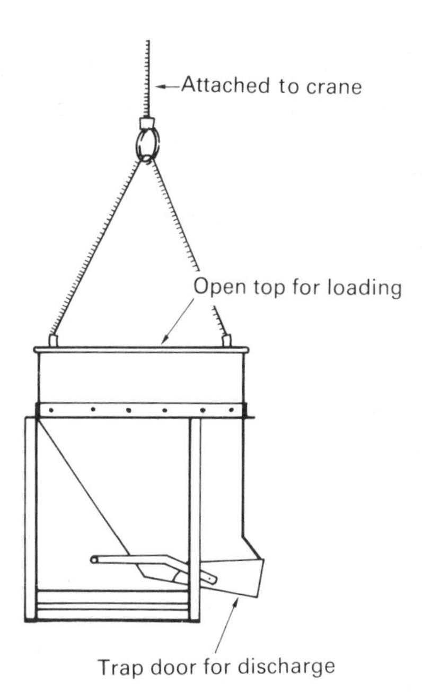

Figure 154 *Vertical movement crane skip*

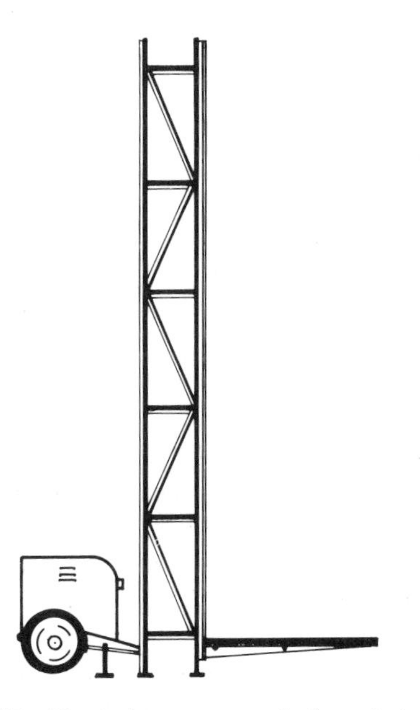

Figure 155 *Vertical movement platform hoist*

Vertical skips are cone shaped and are kept in the upright position by a fixed or detachable frame. This type of skip is often used in conjunction with the detachable skip dumper; loading of the skip is achieved by placing it in an upright position under the discharge point of the mixer.

Where the mixer discharge height is too low to permit the use of this type, a roll-over skip may be used. This has an elongated form which is designed to lie flat on the ground under low discharge height mixers for loading. When being lifted by crane it rolls over into a vertical position.

Both skip types have either a bottom or a side opening trapdoor, which is lever operated or gear wheeled to allow the controlled discharge of concrete into the formwork.

Hoists

The type most often used is a materials hoist consisting of a caged platform on to which a wheelbarrow or pram of concrete is loaded. The platform is raised or lowered to the required level by means of a power unit.

Hoists specifically designed for concrete are also available. These use a skip which is loaded with concrete at ground level. This is elevated to the required working level where it is automatically discharged into a distribution hopper or into a barrow or pram as required for onward transport to the formwork for placing. These distribution hoppers may be equipped with blades or paddles to remix the stored concrete before placing.

Concrete pumps

Concrete pumping is one of the most efficient and quickest methods of transporting and placing concrete. It is particularly suitable for use in conjunction with ready mix concrete on congested sites in inner city areas, where there is insufficient room for other concrete production and transporting plant. Most concrete pumps are hydraulically driven and mounted on either a truck chassis or a trailer unit for mobility. After mixing, the concrete is discharged into the pump hopper from where it is forced by pistons into and along the pipe line. Using fairly small bore pipe lines of 75 mm to 100 mm in diameter, the concrete can be pumped at high speed both horizontally and vertically. The actual output and distances that can be achieved are dependent on the size of pump, the diameter of the delivery pipe, and the mix design.

As an indication, a typical pump performance may be an output of 50 m^3 per hour, with a horizontal pumping distance of up to 500m and a vertical pumping distance of up to 50 m. However, these figures will be greatly affected by the number of bends in the pipework.

The problem of routing the pipework with the minimum of bends etc. can be overcome by the use of pumps with placing booms. These are hydraulically controlled tubular arms with a number of elbow joints on to which the delivery pipework is fixed. The boom can be manually levered both vertically and horizontally by its hydraulic controls to position the discharge end of the pipework over the point of placing.

Concrete placers

These are similar to concrete pumps except that pumps gradually push concrete through the pipeline, whereas placers propel whole batches of concrete at a time along the pipework using compressed air. After mixing, the concrete is discharged into a hopper as with pumping, except that this time an airtight cover is secured down and compressed air is pumped in until the required operating pressure is reached. On opening the delivery valve the air pressure forces the concrete along the pipeline to the point of placing.

Tremie pipes

These are a means of dropping concrete by gravity considerable distances below the source of supply. They greatly reduce the risk of the mix segregating or causing damage to the formwork or reinforcement, which would be almost inevitable if the concrete were permitted to free fall. A tremie consists of a wide topped hopper on to which is attached a metal or plastic pipeline. The hopper permits easy discharge of the concrete into the pipe, whilst the pipeline constricts and guides the concrete steadily into its required place. In addition to their use for placing concrete into deep excavations etc., tremie pipes are also used for underwater concreting.

Compacting concrete

In order to produce a high quality concrete item, it is most important that the concrete is deposited as close as possible to its final position. This is so that it is not moved during its compaction, which can lead to segregation and honeycombing.

Concrete should not discharge from its means of transport to the formwork in one go, so forming heaps which are them moved into position with a poker vibrator. Instead the concrete should be discharged steadily, spreading it evenly as the work proceeds to ensure the minimum possible movement.

The use of a concrete pump or a tremie pipeline when concreting deep walls and columns prevents honeycombing at the bottom of the pour, in addition to the possibility of displacing the steel reinforcing and damaging the formwork. Both can result if concrete is allowed to free fall through reinforcement to the bottom of these forms. Honeycombing or the absence of the cement paste in these circumstances is caused by the paste being left behind on the reinforcement and form sides as it falls downwards.

Compaction of concrete is the process of removing trapped air so as to achieve maximum contact between the constituent materials within a concrete mix. This operation must be commenced as soon as the concrete has been placed, and must be completed quickly before it starts to stiffen. During compaction visible air bubbles rise to the concrete surface. When these cease a thin layer of cement paste or grout will form on the surface. Compaction must be stopped at this time, as over-compaction will result in an excess of grout forming on the surface, producing a weak surface and honeycombing in the lower levels.

Concrete compaction may be divided into three categories – manual, internal vibration and external vibration.

Manual compaction

This method (Figure 156) can only be successfully used for small, relatively unimportant concrete sections where neither high strength nor special finishes are required. The methods used vary. Rodding or tamping with a length of steel bar or timber batten is suitable for lintels, beams and small precast moulds. Tapping the outside of the formwork with a hammer is again suitable for lintels, beams and precast moulds. Tamping the surface with a timber board is suitable for slabs, driveways, paths etc.

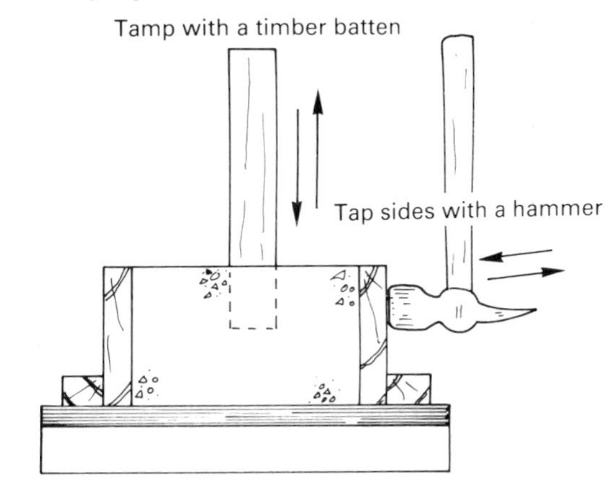

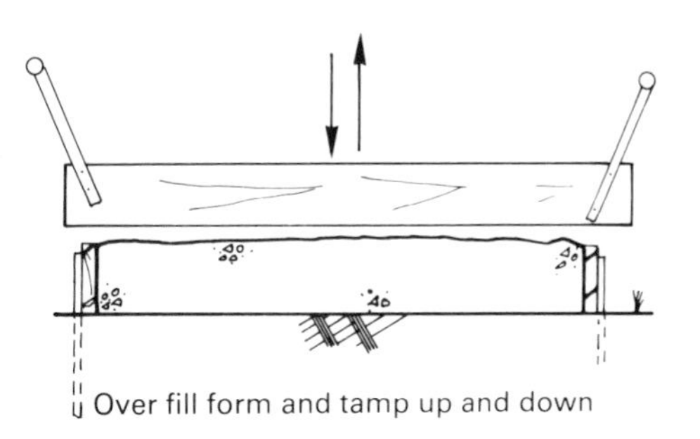

Figure 156 *Manual compaction*

Tamping involves slightly overfilling the form with concrete and them compacting it down by sharply raising and lowering the tamping board down on to the top of the edge forms, working progressively along the slab. Additional concrete can be placed in any hollows. Final finishing is carried out by drawing the tilted board along the form edges over the slab. Compacting by any of the hand methods requires a mix with a fairly high degree of workability, normally with a slump of between 100 mm and 150 mm.

Internal vibration

In most situations this is the most effective method of vibration as the energy is transmitted directly into the concrete (Figure 157). Compaction of the mix is achieved by vibrating the concrete into a fluid state. This releases the trapped air, allowing it to move upwards and escape at the surface. Internal vibrators, or *pokers* as they are commonly known, consist of a vibrating tube about 500 mm long and between 25 mm and 75 mm in diameter. This is connected to its compressed air, electrical or petrol/diesel power supply via a long flexible drive hose.

Concrete for poker vibration should be placed evenly along the forms in layers of between 300 mm and 500 mm deep. The poker is inserted vertically and held in position until air bubbles have ceased rising to the surface. It is then slowly withdrawn and reinserted at intervals of about 250 mm to 500 mm until the whole layer has been covered. Where subsequent layers are being vibrated the poker should penetrate at least 100 mm into the previous one to ensure that the two layers marry together.

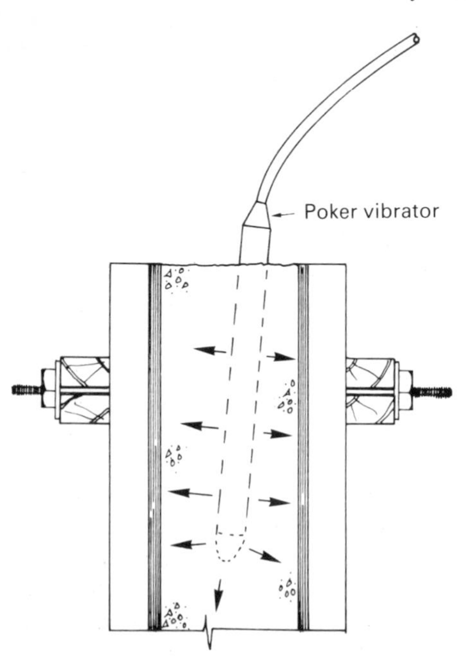

Figure 157 *Internal vibration*

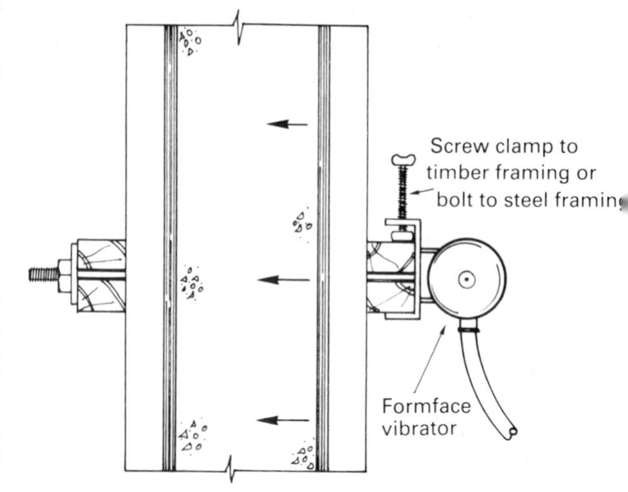

Figure 158 *External vibration*

External vibration

These methods are used in situations where the use of a poker vibrator is impractical e.g. because of the lack of space between steel reinforcement or the narrowness of the section. There are a number of methods (Figure 158).

Surface vibrators

These are used for compacting large surface areas of concrete such as slabs and roads etc. The *vibrating beam* is the most common. It consists of timber or steel beams on to which is mounted a vibrating motor. Its use is very similar to manual tamping in that the level of concrete is overfilled and the beam is moved slowly from one end of the slab to the other on top of the edge forms. This process compacts the concrete, removes surplus concrete and provides a normally acceptable finish. As with hand tamping, any hollow in the surface can be filled by adding extra concrete to the area.

An alternative surface vibrator is the *pan vibrator*. This consists of a steel plate, the edges of which curve slightly upwards so as not to dig into the concrete surface. Attached to the plate is a vibratory unit and a handle. This is guided manually over the surface of the concrete by a single person. Surface vibrators are only really effective up to a depth of about 150 mm or so. Concrete depths in excess of this should be vibrated with a poker before finishing with a surface vibrator.

Form face vibrators
These are used for compacting deep beams and walls. Electric or air driven motors are clamped to the outside of the formwork at intervals. The vibrations are transmitted to the concrete via the formwork. The use of form face vibrators must be considered at the design stage, as the formwork has to be sufficiently strong to withstand the vibrations; for example, timber formwork is best constructed using screws rather than nails, which would loosen during vibration. The motors are required both horizontaly and vertically at about 1 m centres, with alternate rows staggered to ensure complete compaction. In order to minimize on the number of form vibrators required they can be removed and relocated further up the formwork during the pour. To ensure thorough compaction the concrete should be placed in layers no deeper than 150 mm at a time. The upper 500 mm or so of concrete should be compacted either by a poker or manually, as the form vibrator does not effectively vibrate this area.

Hand held hammers
These are driven by compressed air or electric, and may be applied to the form face to supplement other forms of compaction. They are useful to release air trapped against the form face at edges, intricate sections, corners etc.

Table vibrators
These are almost exclusively used in precasting works and concrete laboratories, and only rarely being used on site. They consist of a vibrating spring mounted table, on to the surface of which is clamped the formwork.

Vibrating rollers
These are used to compact dry lean concrete mixes for slab bases, roads etc. Poker or beam vibrators are not suitable for this purpose owing to the dryness of the concrete. Several passes over the surface will be required; with the final pass the roller needs to be in the non-vibrating mode to remove roller marks and seal the surface.

Joints in concrete

Joints are provided in concrete either to allow for subsequent expansion and contraction of material, or to provide a convenient stopping point in the concreting process. These joints are classified as being either movement or construction joints.

Movement joints
Concrete must be expected to expand and contract from its original position (Figure 159).

Expansion joints
These allow for movement occurring as a result of thermal expansion, wetting expansion and subsequent contraction. An expansion joint basically consists of a layer of compressible material (often bitumen impregnated fibreboard) placed between two concrete sections. The exposed edges of the joint will require filling with a waterproof sealing compound to prevent erosion and leakage.

The joint material and its edge sealing compound are designed to compress when the concrete expands, and expand again when the concrete contracts, thus ensuring a gap-free joint between the two concrete sections. Where the concrete sections are reinforced the reinforcement must be stopped short on either side so that it does not pass through the joint.

In order to prevent deflection, load transfer dowel bars are often incorporated in the joint of slabs, roads etc. to transfer the load from one section to the other. In order for the joint to work it is essential that the dowel bar is bonded to one section only and is completely free to move within the other. This is achieved by painting one

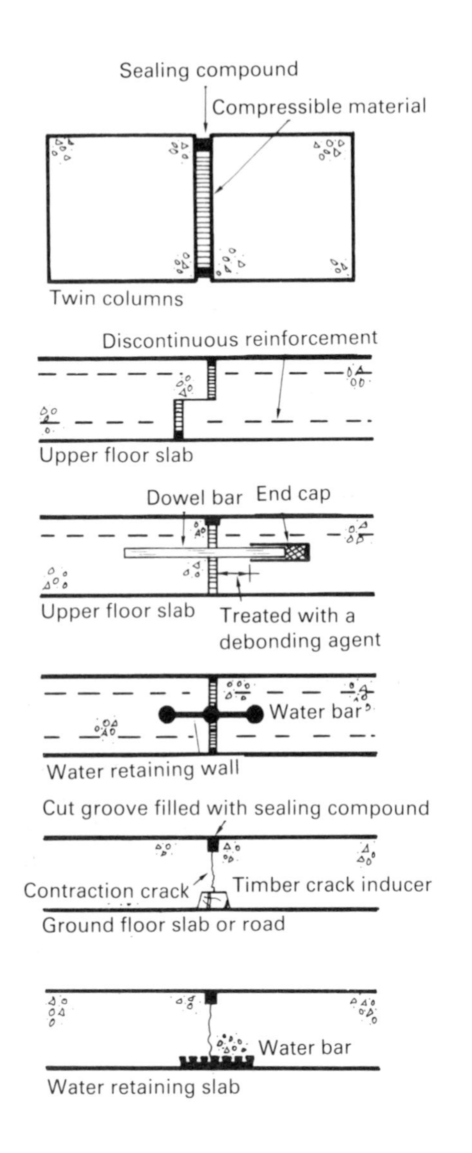

Figure 159 *Movement units*

half of the bar with a debonding agent, and fitting an end cap partially filled with a compressible material.

In water-retaining structures it may be necessary to form a watertight expansion joint. This can be achieved by the inclusion of a proprietary rubber or plastic water bar, which is embedded on either face of the joint or in the centre. To be effective, care must be taken to ensure complete compaction of the concrete around the water bar.

Contraction joints

These permit the natural contraction of concrete as it dries, and also allow for cracking under tensile stress. These joints are formed in predetermined positions by cutting grooves into the surface of freshly hardened concrete to create a line of weakness. On contraction the concrete will crack naturally along the weakened line. Triangular shaped timber crack inducers may be positioned prior to casting along the intended joint to ensure that the crack develops vertically. The cut groove should be filled with a sealing compound to prevent any dirt or grit falling into it.

In common with expansion joints, load transfer dowel bars and water bars may be required in certain situations. Again, for the joint to work effectively any reinforcement must be stopped short on either side of the joint.

Construction joints

Construction joints (Figure 160) are incorporated into a structure:

1 Where it is not possible to complete a concreting operation in one day (hence they are commonly called day joints); or
2 Where it is not practical to cast structural elements simultaneously, e.g. floor slabs and columns or walls; or
3 To allow one section of an element to harden and contract before proceeding with subsequent sections.

The aim with construction joints is to place fresh concrete against hardened concrete and to provide structural continuity as if the joint had not been made in the first place. In order to achieve this it is essential that the joint is correctly prepared.

The hardened surface must be prepared by removing the fairly weak surface layer of cement paste (laitance) and exposing the aggregate. The method used will depend on the age of the

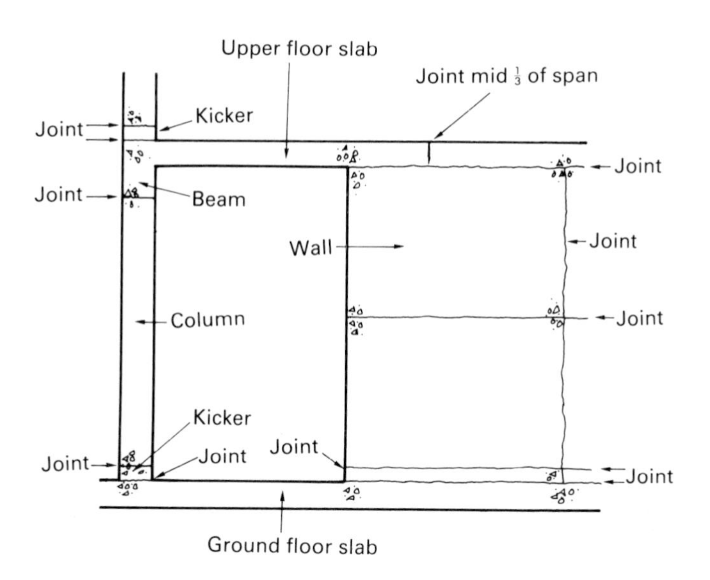

Figure 160 *Typical construction joint locations*

hardened surface. Green concrete may be prepared by removing the laitance with a soft brush and water spray. Hardened laitance may be removed with a wire brush and water spray. Fully hardened laitance will require mechanical tooling with a point chisel or needle gun to prepare the surface. Care must be taken not to shatter or loosen the coarse aggregate (see Figure 161).

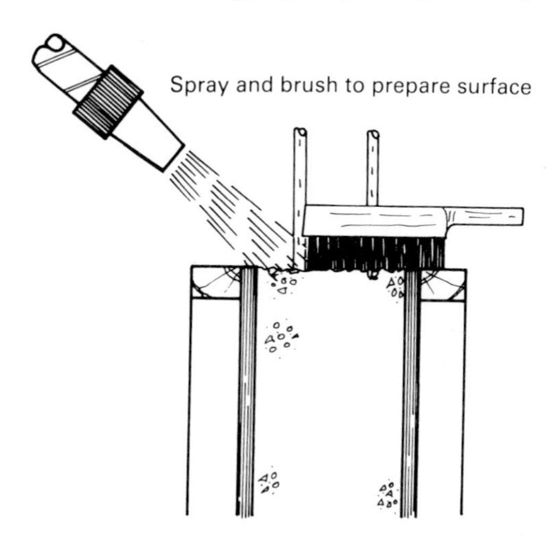

Figure 161 *Preparing a construction joint*

Where construction joints are formed by stop ends in the formwork the faces of these may be treated with a surface retarder. This can be easily removed by a soft brush and water spray after striking to provide an acceptable joint surface.

Construction joints in visible concrete may be masked by making a feature of the joint using suitable formwork details (see Figure 162).

In common with movement joints, water bars may be incorporated in construction joints for water-retaining structures.

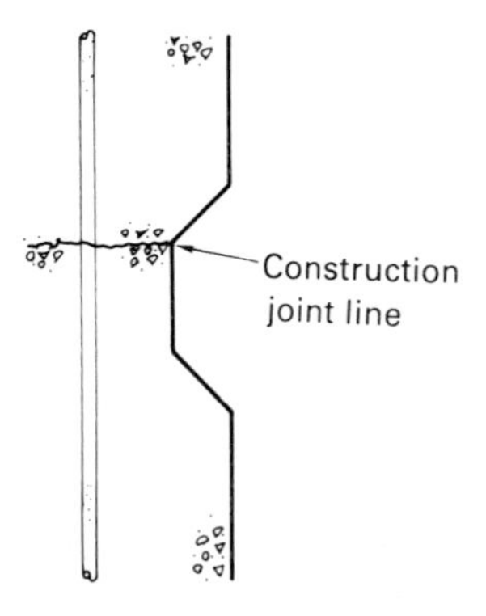

Figure 162 *Use of indent feature to mask a construction joint*

As the joint is structural, the reinforcement should be continuous across the joint and careful compaction of the fresh concrete against the hardened concrete surface is essential. Compressible foam strips are often used to seal the joint between the hardened concrete face and the formwork to prevent grout loss weakening the joint.

Curing

As mentioned previously, concrete gains strength gradually over a period by a chemical reaction called hydration between cement and water, and not (as is often thought) by simply drying out. This hardening process will continue provided that water is present and that the concrete is not subjected to freezing temperatures at an early age.

Curing is the process of retaining water in the concrete and providing insulation from extremes of temperature. A curing period of 7 days (more in winter) from the time of placing is normally considered adequate. Early loss of water will have the effect of reducing the strength that the concrete can achieve. The main causes of this are sun, drying wind, and frost. In addition to allowing the concrete to develop its full strength, effective curing also produces a more waterproof concrete with an increased resistance to abrasion and chemical attack.

Exposure at an early age to either very high temperatures or very low temperatures can result in expansion and cracking of the concrete, before it has gained sufficient strength to withstand these stresses.

Admixtures are useful in assisting the curing process. In hot conditions a retarding agent may be added to slow down the process, whereas in cold conditions an accelerating agent is used which helps to generate the heat of hydration at an early stage.

There are many methods of curing concrete in common use. They can be classified into three groups:

1 Keeping the surface damp
2 Providing an impervious barrier
3 Raising the temperature.

Surface damping

Horizontal surfaces are covered, after the preliminary hardening of the concrete, with wet hessian sacking, wet sand or wet straw (Figure 163). Vertical surfaces such as beams, columns and walls can be kept damp after their forms have been struck by wrapping in several layers of wet hessian sacking. All of the wet covering materials will require spraying with water periodically in order to prevent them drying out, particularly in sunny or windy conditions.

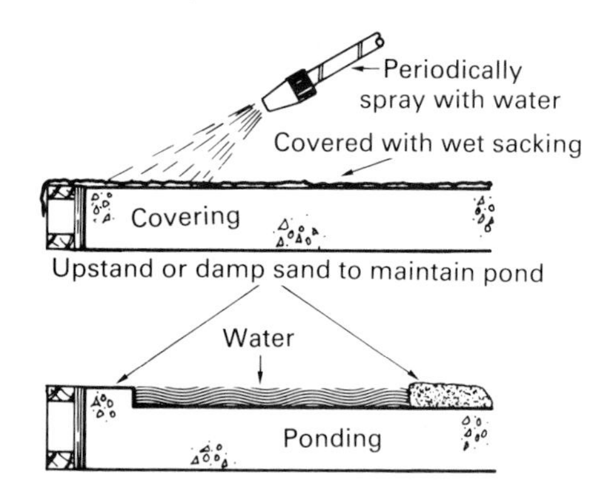

Figure 163 *Curing: keeping surface damp*

Alternatively large horizontal slabs may be kept damp by continuously spraying the surface with water or by ponding. Ponding involves keeping a 50 mm to 100 mm deep pond of water on the concrete surface. The water may be kept in position by an integral concrete upstand beam around the slab perimeter, or by forming a temporary perimeter wall of clay, wet sand etc. Ponding is particularly suitable for curing very deep mass concrete pours, which owing to the very high temperature generated during hydration are very susceptible to cracking.

Clearly any methods of curing that involve keeping the concrete damp will not be suitable for use in freezing conditions.

Impervious barrier

Polythene sheeting, building paper or tarpaulins

are useful for both laying over horizontal surfaces and wrapping around vertical ones (Figure 164). These form an impervious barrier that prevents water evaporation due to strong sun or drying winds. The concrete surface should be sprayed with water before covering, which should take place after the preliminary hardening of the concrete to minimize the risk of damaging the surface. The coverings should extend over the edges of the concrete, and be well tied down to prevent wind lifting them.

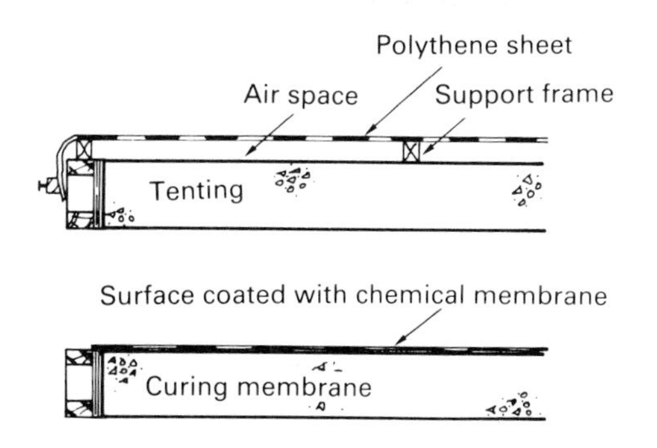

Figure 164 *Curing: covering with an impervious barrier*

Increased resistance to frost can be achieved by tenting the covering 50 mm to 75 mm over the concrete surface on a timber or metal frame. This creates a layer of static air between the covering and the concrete surface, thus providing an insulating layer. To be effective the edges need to be well secured, and any joints lapped by about 300 mm and preferably taped. This not only traps the heat caused by hydration, but also prevents wind from blowing through.

Concrete surfaces can be sprayed with a chemical curing membrane, which forms an impervious layer. The upper surface of concrete should be sprayed at an early age to prevent it being absorbed into the surface pores, causing discoloration as well as reducing its effectiveness. Application just as the glistening sheen of the surface water disappears is preferable. Other surfaces should be sprayed immediately after striking the formwork.

Most of the sprayed curing membranes contain a coloured dye. This aids the spraying operation by helping to ensure a full and even coverage. In hot weather white or aluminized dyes are to be preferred as they reflect much of the sun's heat, thus minimizing any heat gain. Most of these sprayed curing membranes start to wear off after a few weeks exposure to the elements. However, as their impervious layer has the effect of reducing the bond between the concrete and any subsequently applied screed or bonded topping, alternative methods of curing should be considered in these situations.

Temperature raising

In cold weather the curing time can be reduced by raising the temperature (Figure 165). This can be achieved by using a combination of the following. Water may be heated at the mixing stage. Warm air or steam may be blown through the air space between the concrete surface and its tented covering. The surface may be covered with mineral wool insulating blankets (for increased efficiency some are available with electrically heated elements. Space heaters may be used under slabs etc. to raise the temperature, although it may be necessary to temporarily enclose the area by the use of polythene or tarpaulin screens.

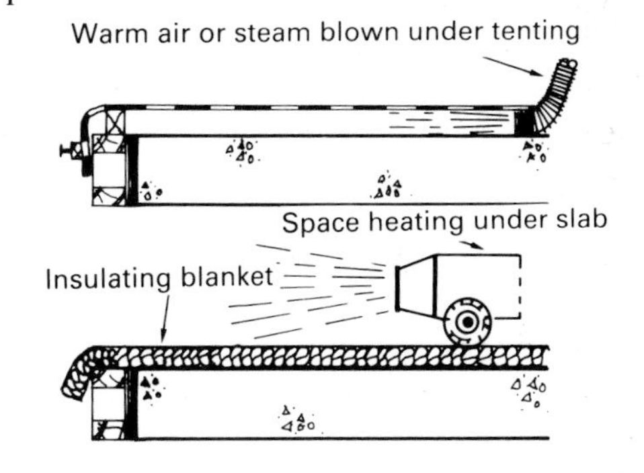

Figure 165 *Curing: raising the temperature*

Self-assessment questions

Question | *Your answer*

1 Briefly describe a method for storing bagged cement on site, and explain the term *first in, first out*.

2 Define the term *batching*, and describe the methods by which this is carried out.

3 State why it is necessary to cover open topped trucks when used for transporting concrete.

4 The purpose of vibrating concrete is to:
(a) Increase the workability to a level suitable for pumping
(b) Enable deep lifts to be poured without risk of segregation
(c) Transfer energy into the concrete to ensure full compaction
(d) Produce an air-entrained concrete mix.

a	b	c	d

5 Distinguish between expansion, contraction and construction joints.

6 The main purpose of curing concrete is to:
(a) Prevent early drying out
(b) Keep it cool
(c) Prevent steel corrosion
(d) Protect from early use.

a [ ]	b [ ]	c [ ]	d [ ]

7 Describe *three* methods used for curing concrete.

8 Explain what is meant by the mixing cycle time.

Part Three

Site Procedures

Chapter 7

Setting out and levelling

Linear measurement

A 30 metre steel tape is commonly used for setting out. The cheaper linen or plastic tapes are not recommended as their tendency to stretch can result in inaccuracies of 6 mm or more in 30 m. Inaccuracies can also occur when using steel tapes if they are not fully stretched out (tensioned). If used in conjunction with a constant tension handle inaccuracies can easily be reduced to less than 3 mm in 30 m.

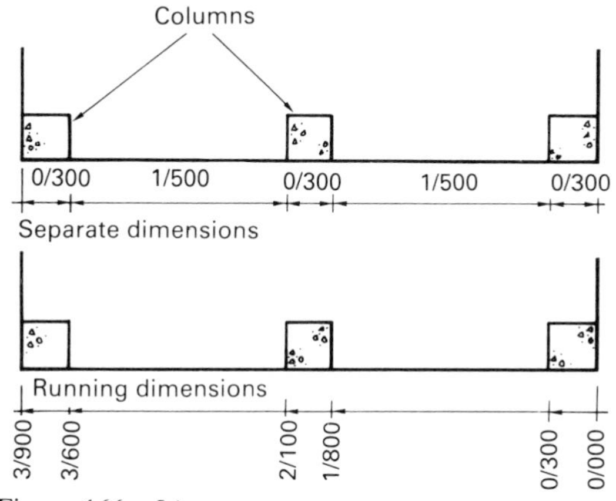

Figure 166 *Linear measurement*

Running dimensions are to be preferred for setting out rather than separate dimensions (see Figure 166), since any inaccuracies or error made in marking one separate dimension will have a cumulative effect, throwing each successive position out. Where running dimensions are not shown on a drawing it is best to work them out and indicate against each position. As a check the total of the separate dimensions should equal the final position figure (see Figure 167).

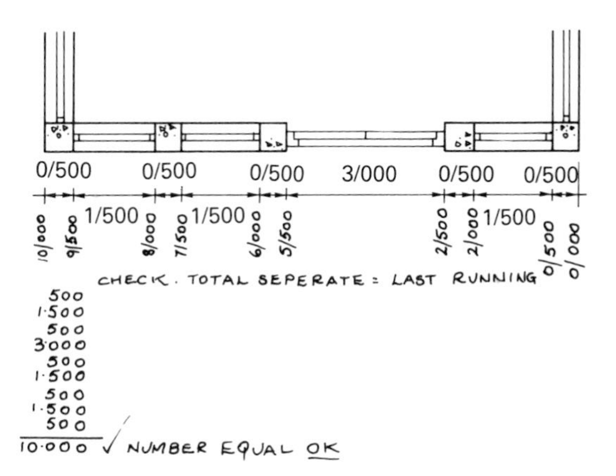

Figure 167 *Marking up a drawing with running dimensions*

Wherever linear measurements are made it is essential that the tape is held horizontal if any degree of accuracy is required.

Setting out angles

Various methods exist for setting out right angles. Three common methods are as follows.

Builder's square

Two of these can be made by cutting a sheet of plywood diagonally across (Figure 168). This is set up on packings so that one edge of the square is against the building line and its corner is adjacent to the right angle position. The side line can be positioned so that it runs parallel to the other edge of the square, thus forming the right angle.

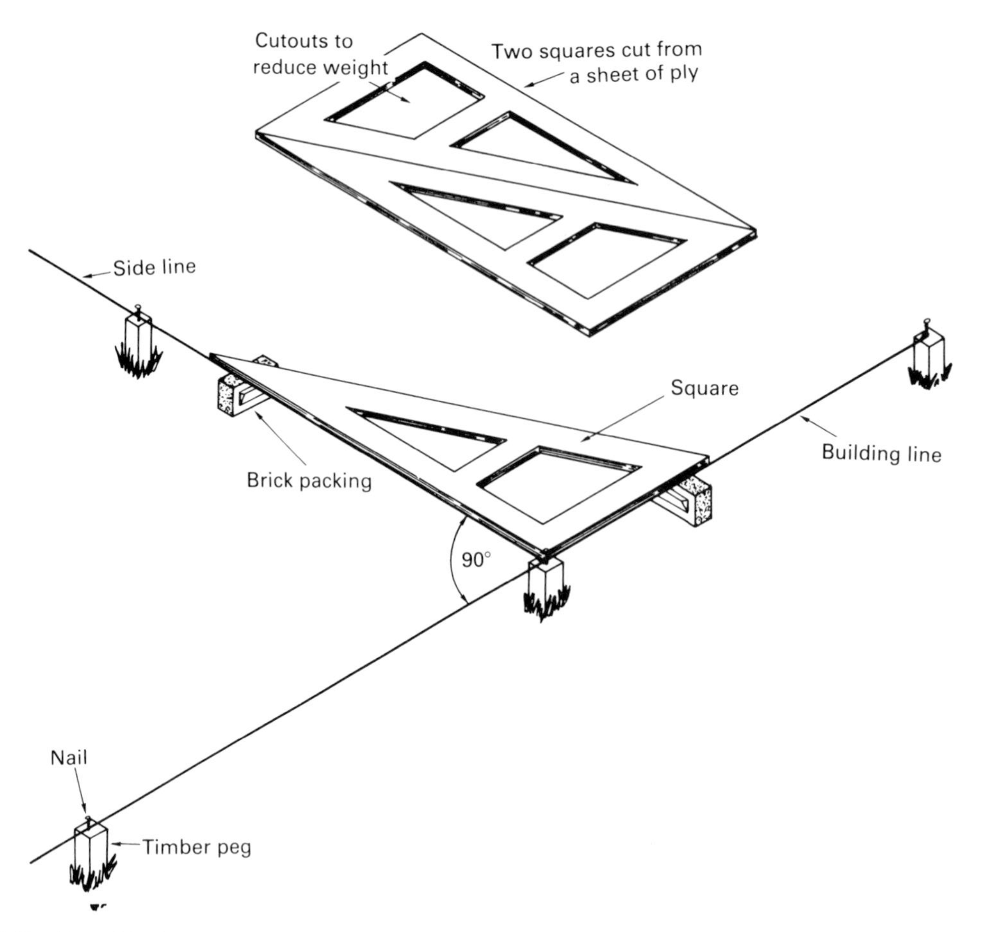

Figure 168 *Use of builders square*

Pythagoras's 3:4:5 rule

This uses the principle that any triangle having sides which measure 3 units, 4 units and 5 units must contain a right angle (Figure 169). Measurements of 3 units and 4 units can be marked out from the right angle position on the front and side lines. The side line is then moved until the distance between the two marked points measures 5 units.

Optical site square

This uses an instrument that basically consists of two telescopes permanently fixed at right angles to each other (Figure 170). The following procedure is used to set out the first right angled corner for a rectangular base:

1. Establish on the building line the front two corners of the base pegs A and B.
2. Set up the instrument over corner peg A.
3. Sight through the lower telescope towards corner peg B.
4. Adjust the fine setting screw and tilt the telescope until the spot on view is achieved.
5. Sight through the upper telescope. Direct your assistant to move peg C into your line of view and drive it partially into the ground.
6. Still sighting peg C, have your assistant drive a

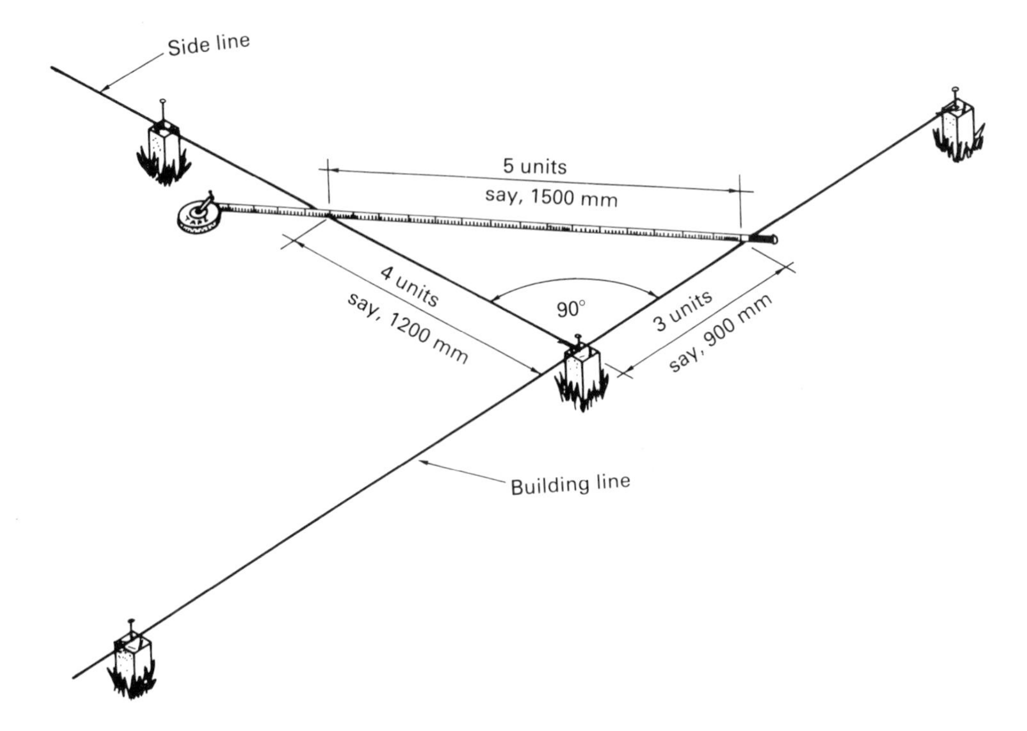

Figure 169 *Use of the 3:4:5 rule to set out a right angle*

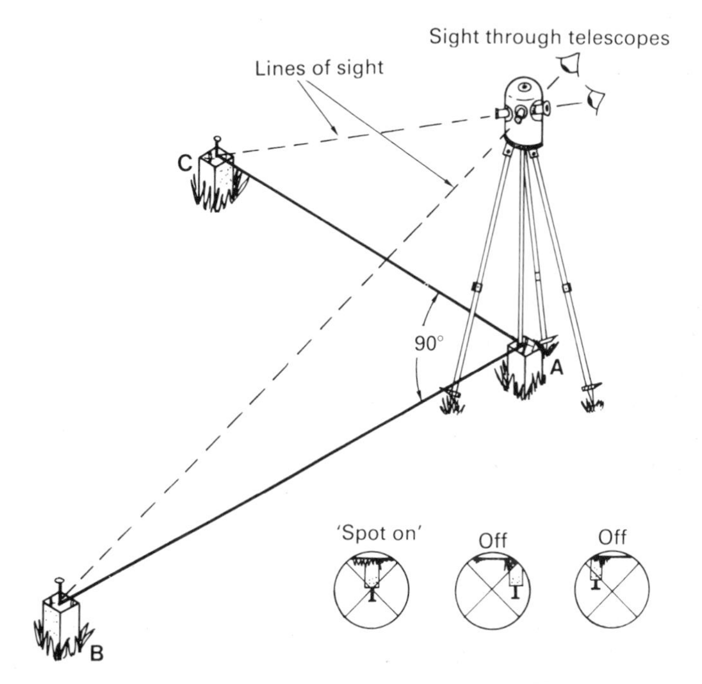

Figure 170 *Use of optical site square*

nail at the spot on position. Angle BAC now forms a right angle.

Setting out curves

The method used for this operation will depend on the size of the curve and its position. Three common methods are as follows.

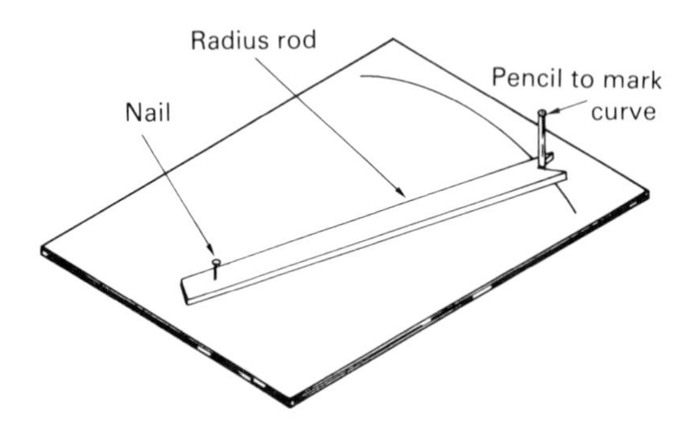

Figure 171 *Use of radius rod*

Radius rod

This is a length of timber with a nail in one end acting as a centre pivot, and a pencil in a vee slot at the other to mark the curve (Figure 171). A radius rod is suitable for accurately setting out curves of up to about 4 m radius.

Triangular frame

This is particularly suitable for large diameter segmental curves or those where the centre is inaccessible. The procedure for making a triangular frame and setting out the curve is as follows (Figure 172):

1 Set out the span and rise of the curve, points A, B and C, and drive a nail into each point.
2 Nail three lightweight pieces of timber together over the points to obtain the triangular frame.
3 Remove the nail at point B.

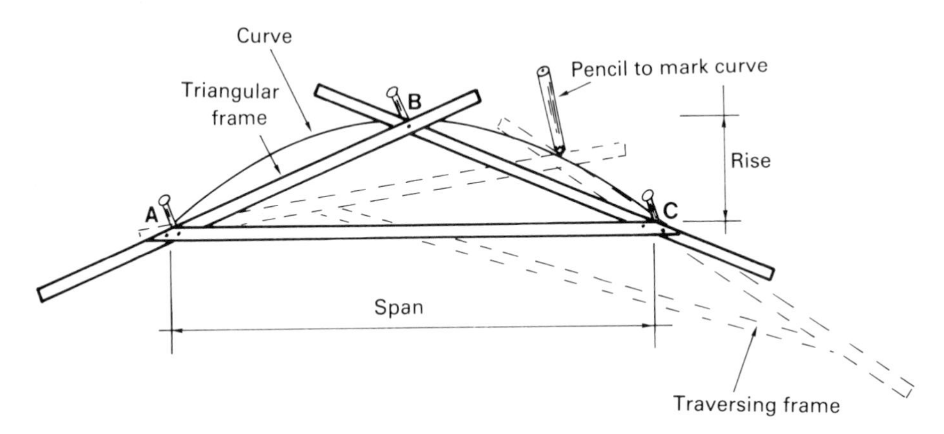

Figure 172 *Triangular frame*

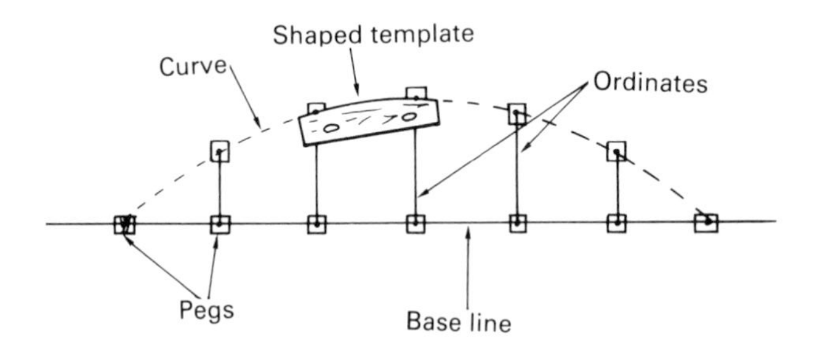

Figure 173 *Calculated ordinates*

4 Move the frame from side to side, keeping it in contact with the other two nails, using a pencil at the apex to mark the required curve.

Calculated ordinates

This method is useful for setting out very large radius curves, where the physical size of a radius rod or a triangular frame would be awkward to handle (Figure 173).

The procedure is to establish a base line; calculate the length of a number of ordinates; peg these out at right angles to the baseline; and use a shaped template between the ordinate pegs to mark the smooth line of the curve.

Pythagoras's theorem is used when calculating the chord length. Where the radius length is unknown, this can first be calculated using the intersecting chord rule.

Example

To calculate the ordinate lengths required to set out a curve having a chord length (base line) of 15 m and a rise at midpoint of 1.5 m (Figure 174).

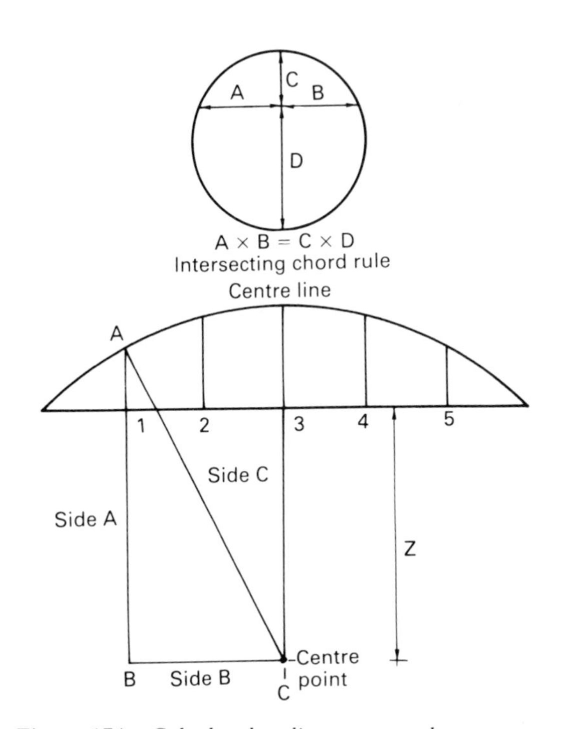

Figure 174 *Calculated ordinates example*

Use the intersecting chord rule to calculate radius:

$$A \times B = C \times D$$

$$D = \frac{A \times B}{C}$$

$$= \frac{7.5 \times 7.5}{1.5} = 37.5$$

Therefore

$$\text{radius} = \frac{D + C(\text{diameter})}{2}$$

$$= \frac{37.5 + 1.5}{2} = 19.5$$

Divide the chord length into an even number of parts, say six:

$$\frac{15}{6} = 2.5$$

This gives ordinates 1 to 5 spaced 2.5 m apart. Only ordinates 1 and 2 require calculation; ordinate 3 is 1.5 m, and 4 and 5 will be the same lengths as 1 and 2.

Use Pythagoras's theorem, i.e.

$$A^2 + B^2 = C^2$$

(where C is the longest side of a right angle triangle), to calculate the ordinate lengths.

Draw a right angled triangle ABC, where:

AC = radius
BC = measurement along chord from centre line to ordinate
AB = unknown ordinate 1

Then

$$A^2 + B^2 = C^2$$
$$A^2 = C^2 - B^2$$
$$A^2 = 19.5^2 - 5^2 = 355.25$$
$$A = \sqrt{355.25} = 18.848 \text{ m}$$

Therefore:

$$\text{ordinate } 1 = 18.848 \text{ m} - Z$$

where

$$Z = \text{radius} - \text{rise on centre line}$$

$$Z = 19.5 - 1.5 = 18 \text{ m}$$

Hence

ordinate 1 = 18.848 − 18 = 0.848 m = 848 mm

Repeat the procedure to calculate the length of ordinate 2:

$$A^2 = C^2 - B^2$$
$$A^2 = 19.5^2 - 2.5^2 = 374$$
$$A = \sqrt{374} = 19.339\ m$$

Therefore:

$$\textit{ordinate 2} = 19.339 - Z$$
$$= 19.339 - 18 = 1.339\ \text{m} = 1339\ \text{mm}$$

Setting out a building

Small buildings

To establish the position of a small building, such as the concrete base for a small warehouse, the following procedure can be used (see Figure 175). All setting out is done from the building line (normally the front line of the building). This will be shown on the block plan. Its position is determined by the local authority.

1 Establish the building line by positioning pegs A and B on the side boundaries. Nails are driven into the top of the pegs to indicate the exact position, and a line is strained between them.

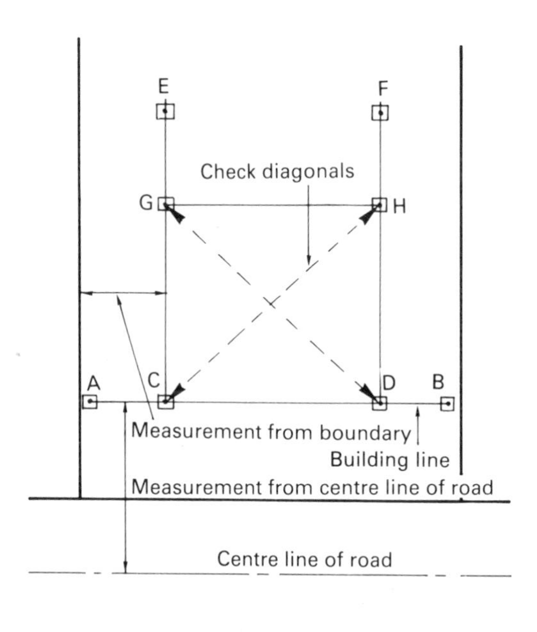

Figure 175 *Setting out a building*

2 Position pegs C and D along the building line to indicate the two front corners of the building. The distance between them and their position in relation to the side boundaries will again be given on the architect's drawings.
3 Establish lines from pegs C and D at right angles to the building line to position pegs E and F using either of the methods previously covered. Strain lines between pegs C and E and pegs D and F to indicate the two sides of the building.
4 Measure from the building line along lines CE and DF to establish pegs G and H in the far corners of the building.
5 Measure the diagonals CH and DG. If the setting out is accurate these dimensions should be the same. If not, a check through the previous stages must be made to discover and rectify the inaccuracy.
6 Set up the profile clear of the pegs at the four corners as shown in Figure 176. Transfer the positions of the setting out lines on to the profile boards (see Figure 177).
7 The setting out lines and pegs can now be removed. These are replaced by lines strained between the nails on the profiles. Where these lines cross are the corner positions of the base. Plumb down from these lines to position the formwork (see Figure 178).

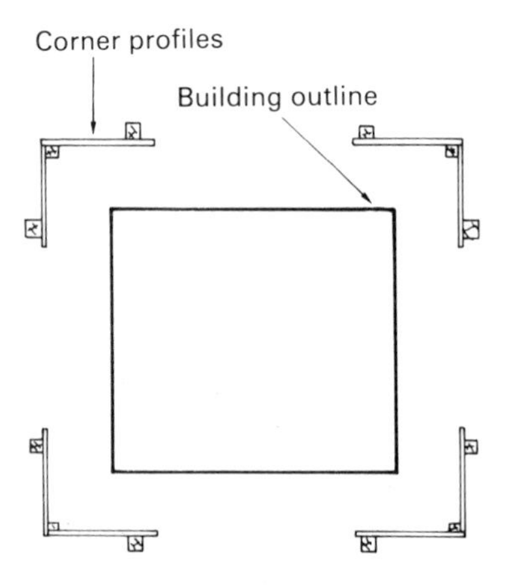

Figure 176 *Profile boards*

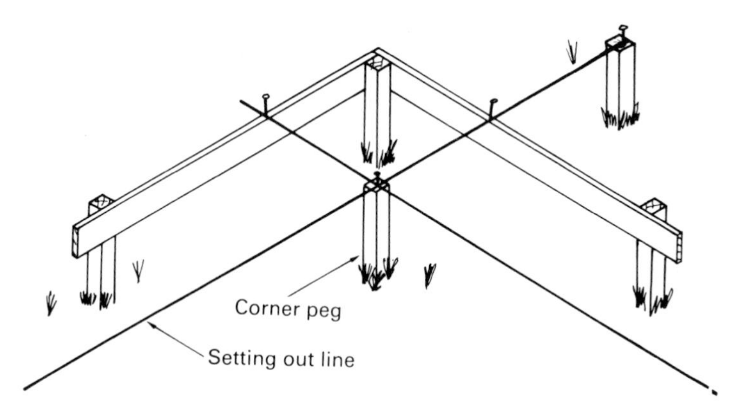

Figure 177 *Transferring setting out lines to profile boards*

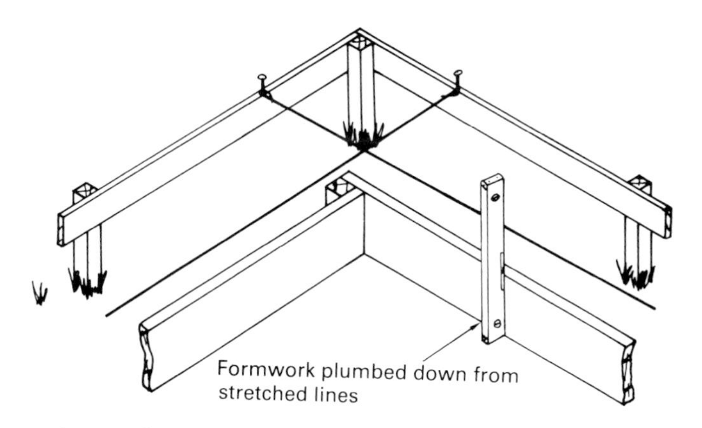

Figure 178 *Positioning formwork*

Skeleton framed buildings
Skeleton framed buildings are set out using a structural grid. A typical example is illustrated in Figure 179. The basis of the grid is formed using the intersecting centre lines of the columns, walls and beams. In addition a reference system which identifies each structural element can be created by lettering the centre lines in one direction and numbering them in the other.

To establish the previous structural grid, the following procedures can be used:

1 Establish, using the previous procedure, the four corner pegs of the building.
2 Position profiles along each side of the building. These may need to be well clear of the work if access is required for excavators etc.
3 Transfer the corner points on to the profiles. From these corner points measure out the centre line positions of the columns, walls or beams.
4 Drive nails along the profiles and strain lines between them to form the centre line structural grid.
5 Centre lines for positioning formwork can be obtained by plumbing down from the grid (see Figure 180).

Establishing levels

Levels on site to set formwork at the correct height for column bases and beams etc. are normally transferred from a datum peg (also

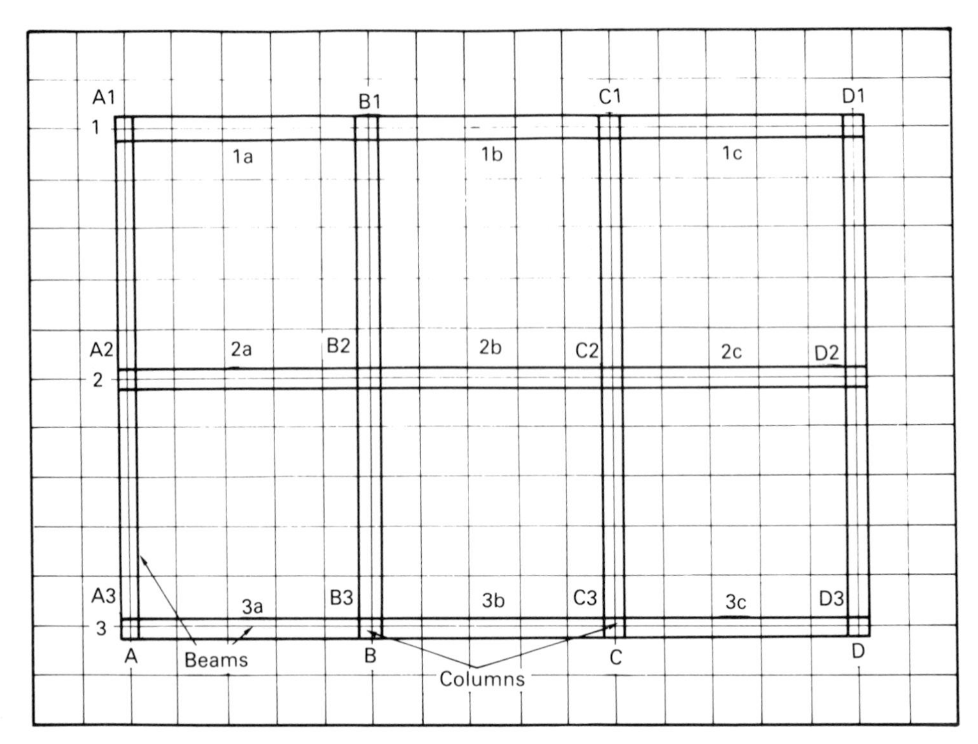

Figure 179 *Structural grid*

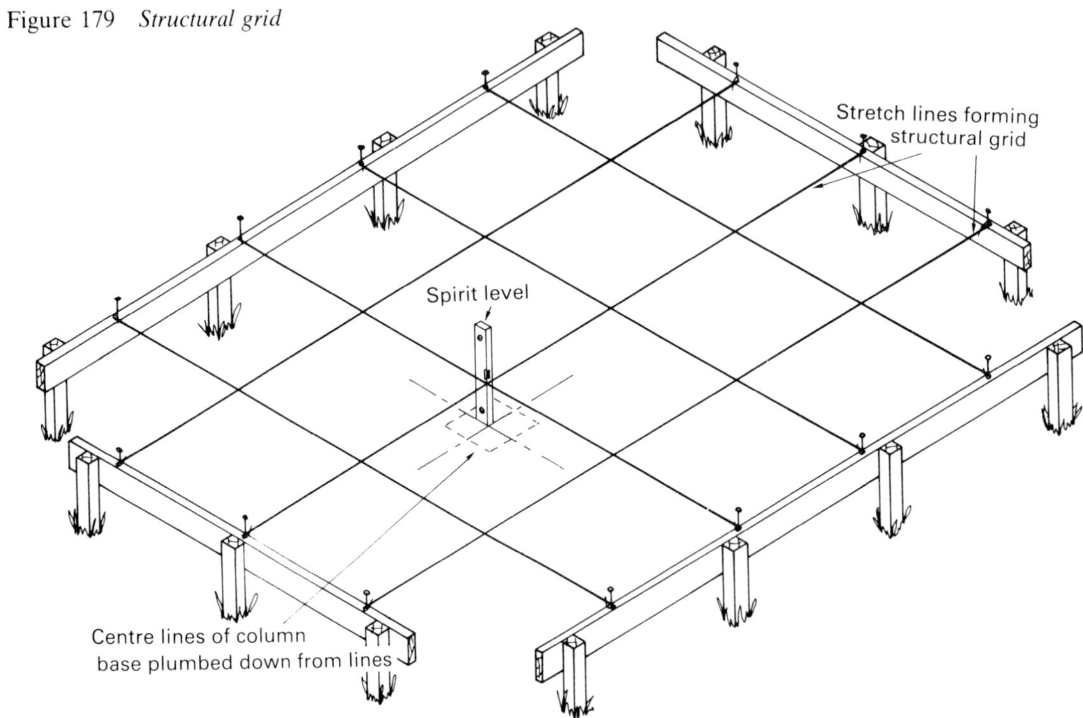

Figure 180 *Use of a structural grid*

known as a temporary bench mark (TBM)) using a quickset tilting level.

The level value of a datum relates to the Ordnance Bench Mark (OBM) from which it was established. The OBM itself will be a set height above the Ordnance datum, which is the mean sea level at Newlyn in Cornwall.

Illustrated in Figure 181 is a typical site datum peg. This is either a timber peg or a steel pin driven into the ground to a suitable level and then set in concrete for protection. Also shown is an OBM; these are to be found cut into the external walls of churches, schools and other public buildings, normally close to ground level. The level value of OBMs can be obtained from the relevant Ordnance Survey map or from the local authority's planning office. If the value obtained for a particular OBM is say 50.000, this means that the centre of the horizontal bar is exactly 50 m above the Ordnance datum.

Use of a quickset tilting level

A quickset tilting level consists basically of a telescope, with a spirit level fixed to its side to ensure a level line of sight (see Figure 182). It is attached to a tripod by means of a ball and socket joint. Levels are established with the aid of a graduated measuring staff.

Telescopic staffs in either timber or aluminium are most often used. They consist of a number of sections that slide within each other. Each section is graduated to permit direct readings in metres, 100 mm and 10 mm divisions. Measurements falling between the 10 mm divisions are read by estimation.

Figure 183 illustrates a portion of a typical staff with various readings indicated. Alongside this is shown an inverted image of the staff as seen through most levelling instruments. (Some levels are equipped optically to permit readings the right way up.) The actual level reading is taken on the centre cross-hair lines.

Setting up a quickset level

Open up the tripod, extend its legs if required, so that the instrument will be approximately at eye level; ensure stability by treading the steps at the bottom of each leg firmly into the ground. Ensure all wing nuts and fittings are tight. Place the instrument on the tripod, locate the screw thread and tighten once the circular bubble has been centralized.

Focus the cross-hairs by adjusting the eyepiece. Place the palm of the hand just forward of the telescope; this directs a uniform amount of light into the instrument. Sight through the eyepiece; at the same time rotate it until the cross-hairs are seen as sharp and clear as possible.

Focus the instrument by adjusting the focusing screw. Sight the staff through the eyepiece; at the same time adjust the focusing screw until the staff

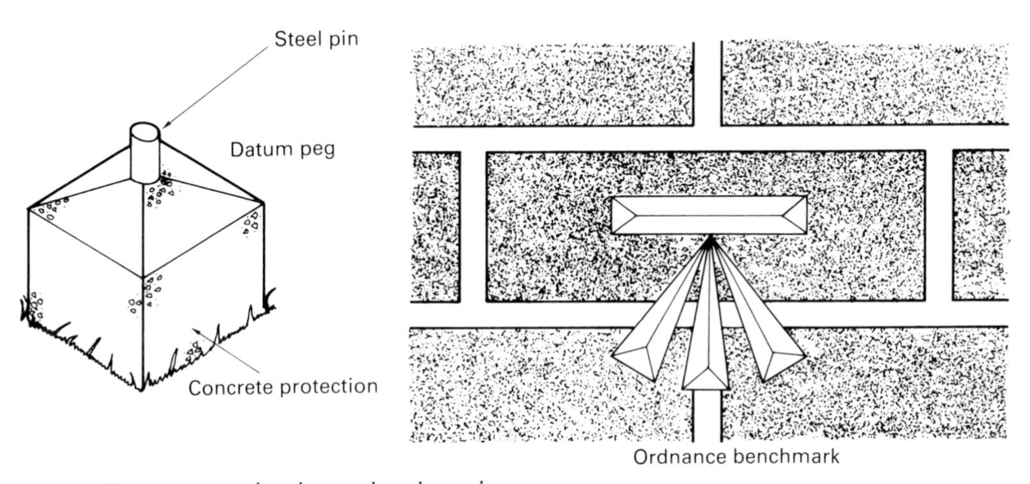

Figure 181 *Datum peg and ordnance bench mark*

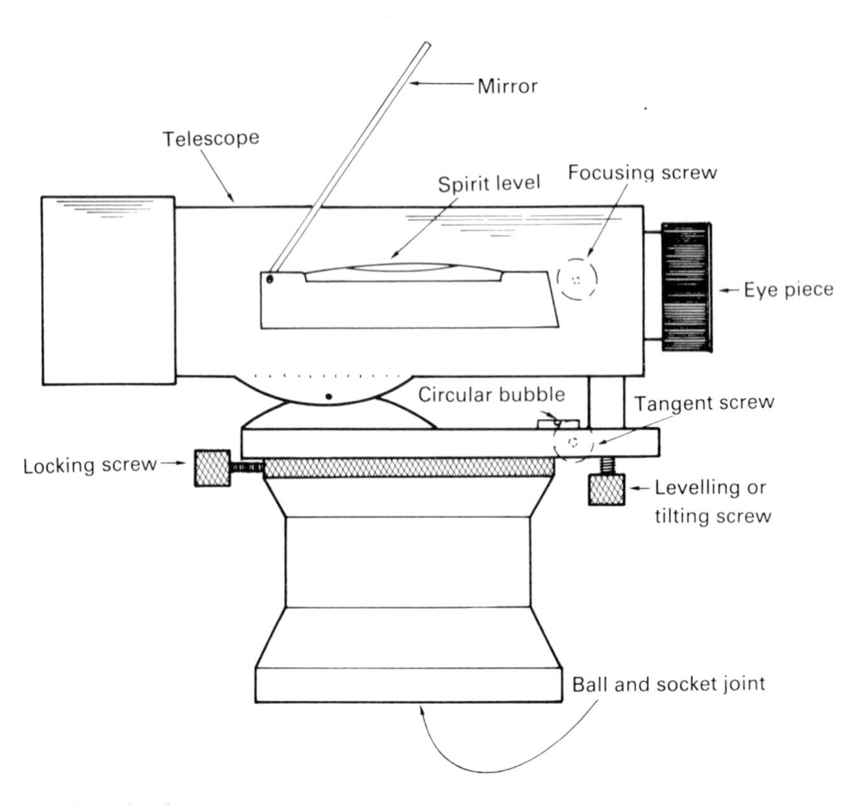

Figure 182 *Quickset tilting level*

is seen sharp and clear. Adjust the tangent screw so that the staff is viewed central.

Errors in readings due to parallax can occur if the cross-hairs and focus are incorrectly adjusted. Parallax is the apparent separation between the cross-hairs and the staff image. If the instrument has been incorrectly set up, parallax will be noticed by the user as small changes in the staff reading as the eyes are moved up and down. If this is the case, readjustment is required. This procedure should be used as a check each time the telescope is focused onto a new staff position.

Rotate the tilting screw so that the bubble in the spirit level is central when viewed in the mirror. It is essential that the bubble is centralized each time a reading is taken, otherwise the result will be inaccurate, as the line of sight may not be truly horizontal.

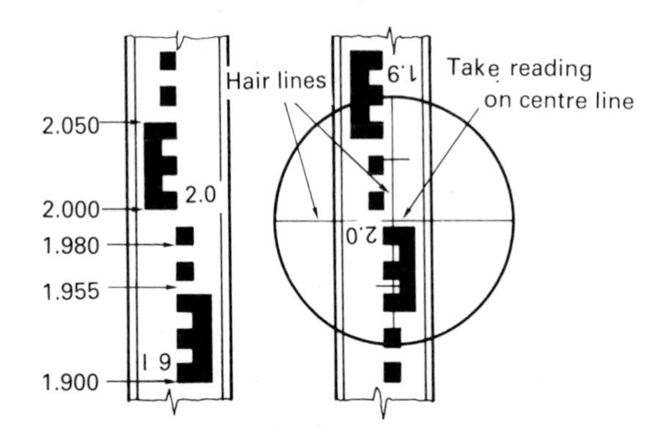

Figure 183 *Staff readings*

Using a quickset level

The following basic terminology is commonly used when transferring and recording levels:

Backsight The first reading taken after setting up the instrument.

Foresight The last reading taken before moving the instrument to a new position.

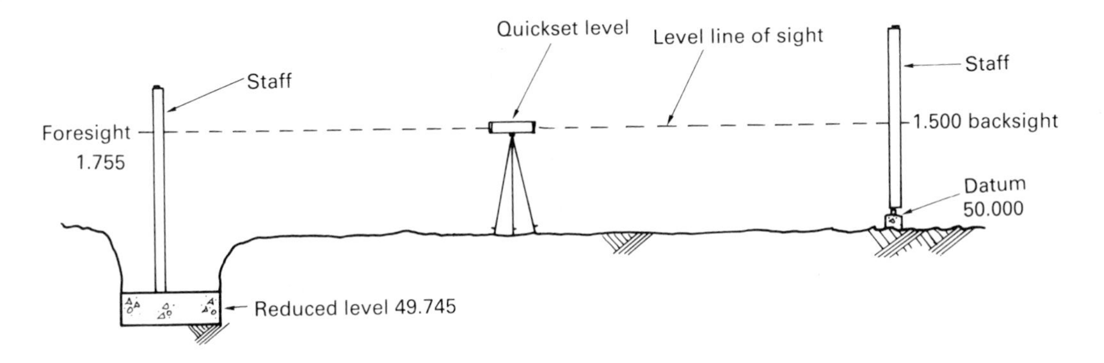

Figure 184 *Using a quickset level*

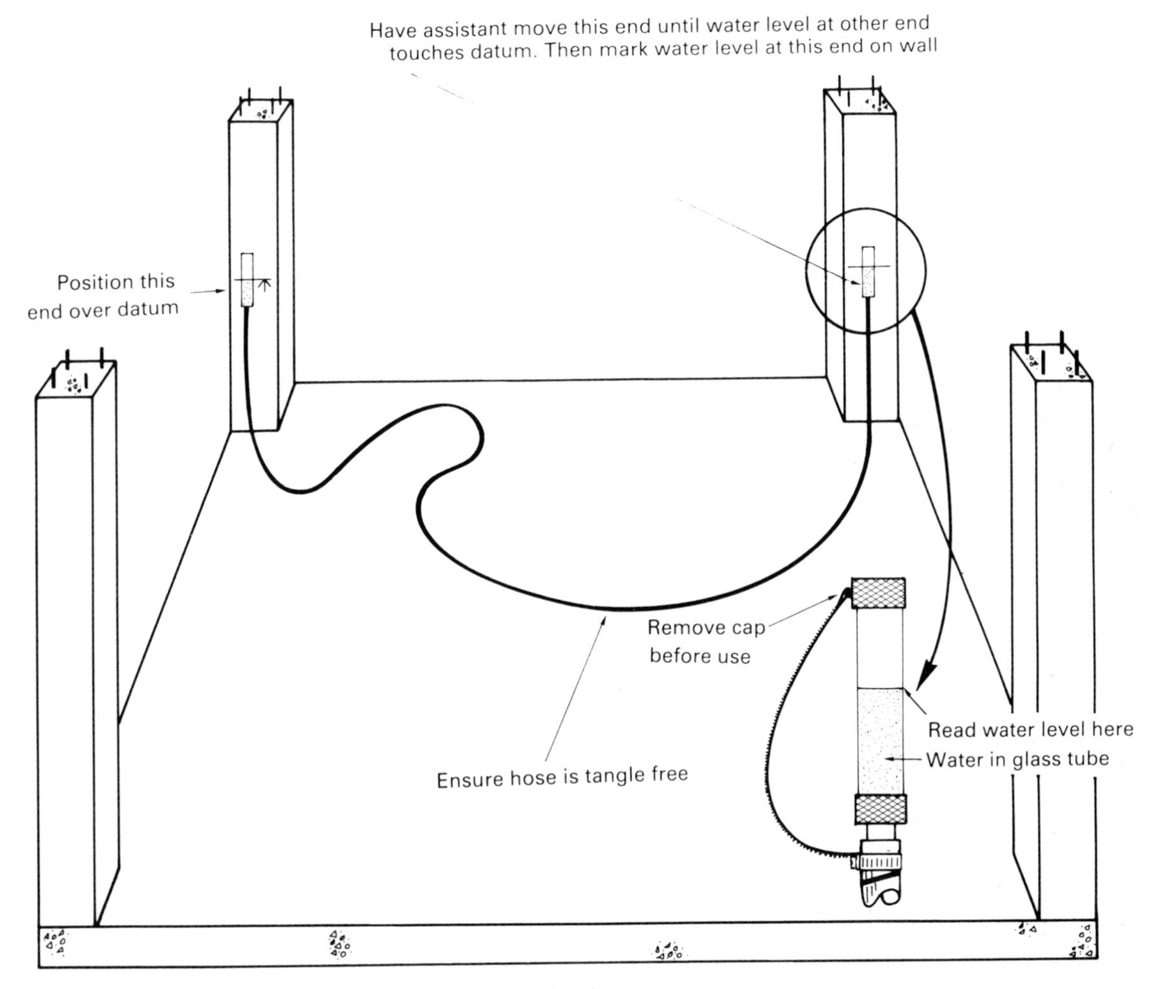

Figure 185 *Transferring datum marks with a water level*

Intermediate sight All other readings taken between the backsight and the foresight.
Reduced level The height of any position above the Ordnance datum.
Height of collimation The height of the line of sight of an instrument above the Ordnance datum.

The procedure for taking and recording levels using the height of collimation method is as shown in the following example.

Example
To determine the level value (reduced level) of a concrete base (see Figure 184).
The reduced level of the base will equal the height of collimation minus the staff reading at the reduced level. The height of collimation equals the level value of the datum plus the staff reading at the datum.

The procedure is as follows:

1 Set up the instrument ready for use midway between the datum and the concrete base.
2 Take staff reading at datum: 1.500. This is a backsight and is recorded in the backsight column of Table 22. Since the reading is taken at the datum its reduced level of 50.000 is recorded in the reduced level column. The backsight and the reduced level are added together to give the height of collimation, 51.500, which is recorded in the appropriate column.
3 Resight the staff with it placed on the concrete base: 1.755. As this is the last reading, it is recorded in the foresight column.
4 Determine the reduced level and record it in the appropriate column:

   reduced level = height of collimation − staff reading

   = 51.500 − 1.755 = 49.745
5 As a check for both recording and calculation errors, the difference between the backsight and the foresight should equal the difference between the first and last reduced level (FRL and LRL). This check is shown in Table 22.

It should be noted that the reduced level of any point can be determined by subtracting the staff reading from the height of collimation. Conversely the staff reading required to set any reduced level can be determined by subtracting the reduced level from the height of collimation.

Example
Determine the reduced level of the top of a strip foundation, if

datum = 45.000
staff reading on datum = 1.250
staff reading on strip foundation = 1.805

Reduced level

= height of collimation (datum + staff reading on datum) − staff reading on strip foundation
= (45.000 + 1.250) − 1.805 = 44.445

Therefore the reduced level of the top of the strip foundation is 44.445 m above the Ordnance datum.

Table 22

Backsight (BS)	*Intermediate sight*	*Foresight (FS)*	*Height of collimation*	*Reduced level (RL)*	*Notes*
1.500			51.500	50.000	From
		1.755		49.745	datum
Check:	FS	1.755	FRL	50.000	to base
	BS	1.500	LRL	49.745	RL
		0.255		0.255	
Differences equal OK					

Example
Determine the staff reading required to set the top of the formwork sides to a ground beam at 98.475 if

datum = 100.000
staff reading on datum = 1.550

staff reading required

= height of collimation − reduced level
= (100.000 + 1.550) − 98.475 = 3.075

Thus the staff reading required is 3.075 m.

Datum marks

Datum marks or lines are positions marked on walls and columns of a building in order to provide a reference point for positioning other building elements and components.

A datum mark is established at a convenient height, say 1 m above the finished floor level (FFL) and marked thus:

⩚ an arrow indicating a horizontal level line

To establish datum marks on the columns of a framed building, mark the datum at a convenient height on the first column. Transfer this to the others in turn using either a water level as shown in Figure 185 or a quickset level as shown in Figure 186.

Note Before using a water level it must be

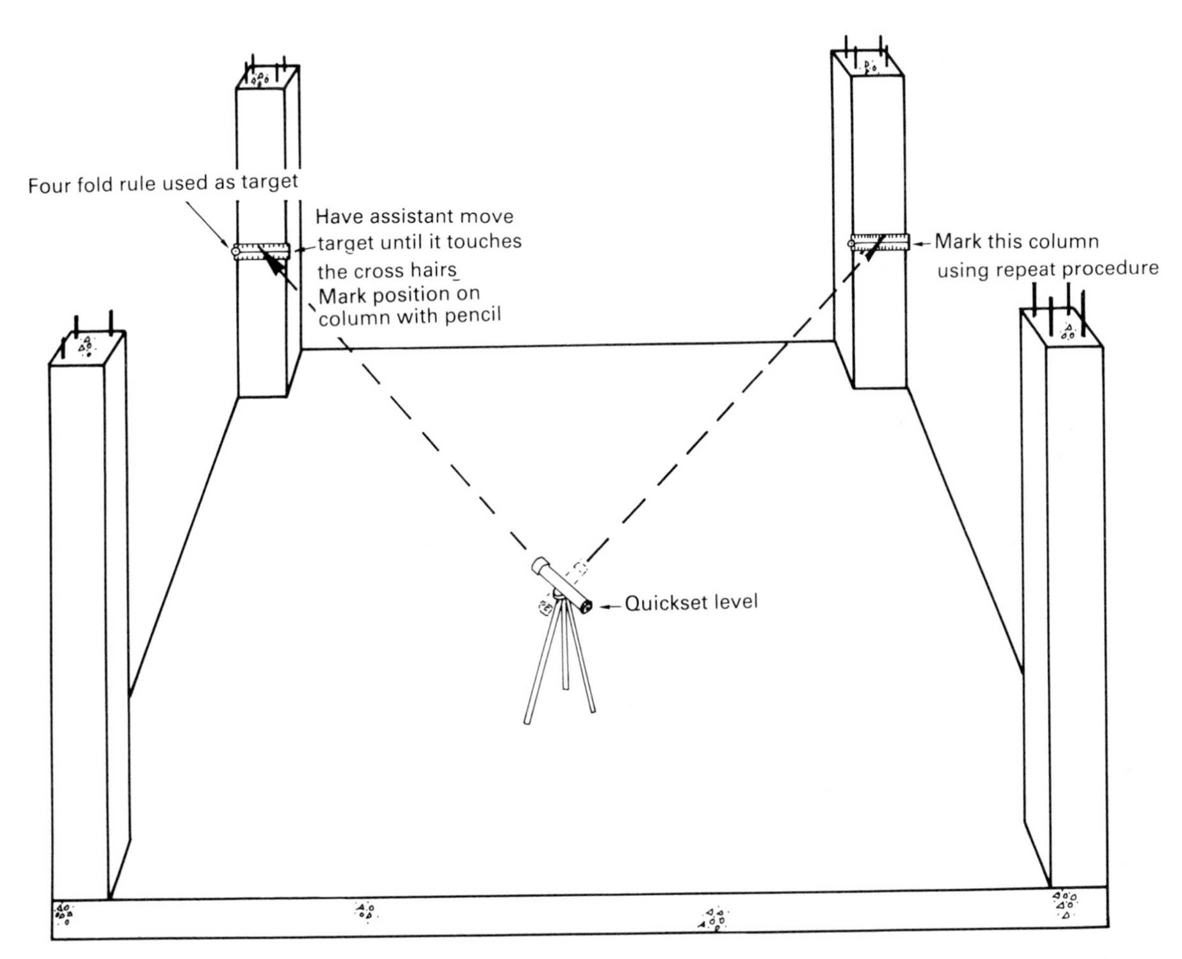

Figure 186 *Transferring datum marks with a quickset level*

prepared by filling it from one end with water, taking care not to trap air bubbles. Check by holding up the two glass tubes side by side: the levels of the water should settle to the same height.

The positions on the columns can then be used as a level reference point. For example the underside of the slab (decking level) can be marked on the top of each column for positioning the formwork by measuring up a set distance from each datum.

Having erected the slab formwork to the correct height, it should be checked for level and adjusted as required. This operation can be carried out with the aid of a quickset level using the following procedure (see Figure 187):

1 Set the instrument up and adjust ready for use, adjacent to the slab formwork and at the same level. This may be either on a previously cast section of the same slab or on an adjacent scaffold. Never set up the level on the actual formwork as the instrument will move along with the formwork as it is adjusted.
2 Two assistants are required, one on the decking holding the staff, the other below the decking to adjust the props. It is unlikely that the person below will hear verbal instructions, so prearrange a code system, e.g. one stamp with the foot on the decking to wind up the prop, two stamps to lower it.
3 The levels are checked along the lines of each ledger, since it is these that support the joists and decking.
4 Starting at the ledger adjacent to the columns or wall, adjust the props until the top of the decking coincides with the previously established mark. Take a staff reading at either end and midway along this ledger. Each reading should be the same. Record this, as it is the reading required to level the whole deck.
5 Position the staff at one end of the next

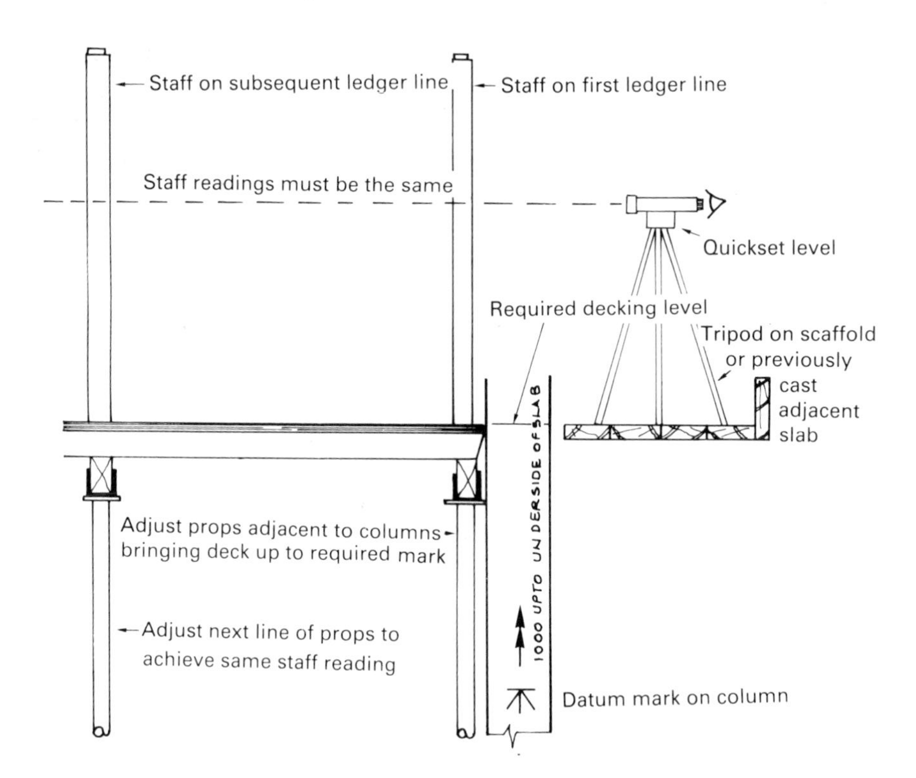

Figure 187 *Using a quickset level to adjust and level slab formwork*

ledger; adjust the prop until the same reading is achieved. Reposition the staff, first at the other end of the ledger, followed by midway along; adjust props as required to achieve the desired reading. Long ledgers or those formed in two lengths will require additional intermediate checks.

6 Repeat the previous stage for each remaining ledger.

After the levelling operation is complete and before loading the formwork, check to ensure that all the prop collars are adjusted tight up to the pin, as the process of adjusting various props can cause the inner tubes of intermediate props to lift up. If the collars were not so adjusted the formwork would sag and distort under load.

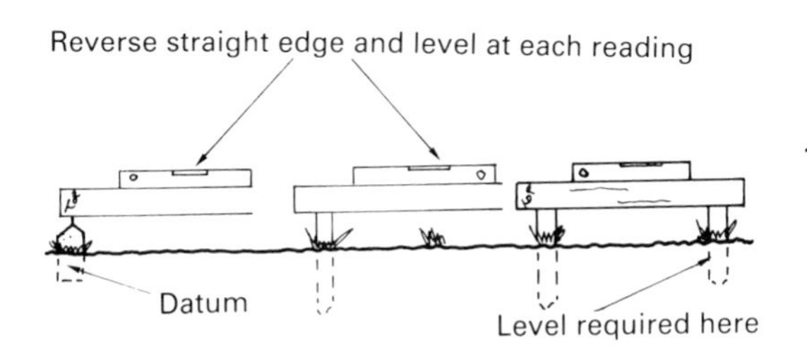

Figure 188 *Use of a straight edge and spirit level*

Simple levelling operations may be carried out using a straight edge and level. To obtain maximum accuracy and not multiply any slight error in the level or unparallel straight edge, it is recommended that both the level and straight edge are turned end to end each successive time the level is transferred (see Figure 188). As an illustration of the possible inaccuracies, if a level and a 3 m long straight edge with a 5 mm inaccuracy were used to transfer a level over 12 m without turning, it could be ±20 mm out of level. However, if used correctly the level would be spot on.

Verticality

The verticality or plumb of a building can be simply achieved using a spirit level or a suspended plumb bob. More specialized methods often employed on larger structures include the use of theodolites, laser beams and optical auto-plumbs, etc. As far as the formworker is concerned the verticality of wall and column formwork is normally checked using either a spirit level and straight edge with leg attached to clear clamps, walings etc.; or a plumb bob suspended clear of the form, with measurements taken at the top and bottom.

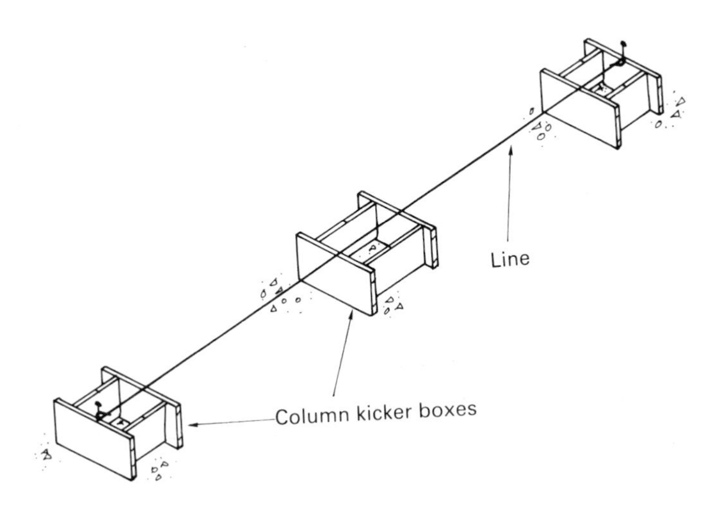

Figure 189 *Lining in column kickers*

Line

In addition to being level and plumb, form faces must also be in line. You have previously covered how the centre lines of column bases and beams etc. are set out from a structural grid. A similar procedure can be used to check the line of both column kickers and erected column boxes. If the centre lines are marked on the form's top edges, a line can be strained between the end boxes and the intermediate ones can be positioned accordingly. Figure 189 illustrates this method of lining column kicker boxes.

Wall forms, slab edges, beam sides and the like can be 'lined in' using the method illustrated in Figure 190. This entails plumbing and fixing the two extreme ends. Drive nails near the top of the two ends and strain a line between them. Cut three identical pieces of packing, say 18 mm plywood; place one each end under the line. Use the third piece to check the distance between the form face and the line at each support position. Adjust the prop or strut so that the packing piece just fits between the form face and the line.

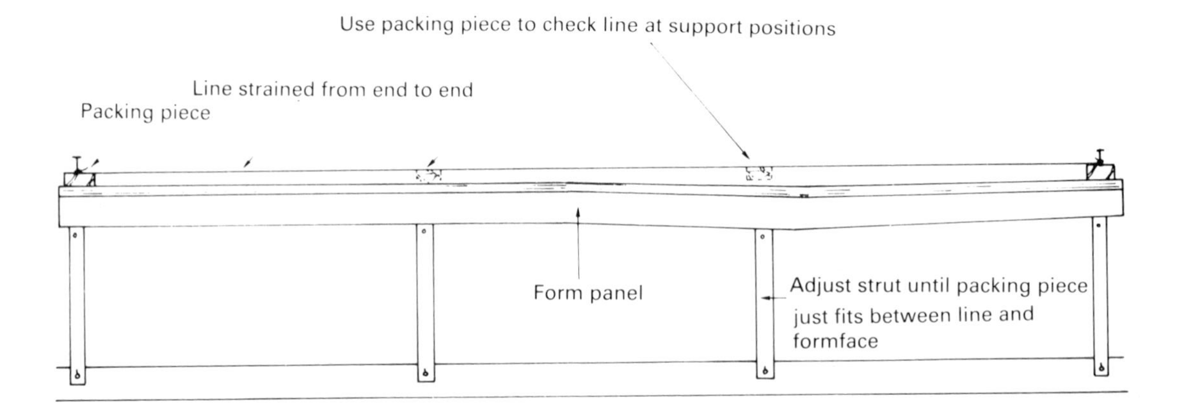

Figure 190 *Lining in*

Self-assessment questions

Question | *Your answer*

1 Determine the staff reading required to establish a reduced level of 46.250 for the formwork to a column pad base, if a reading of 1.550 is recorded with the staff set on the 50.000 TBM.

2 Explain the difference between separate and running dimensions shown on a plan drawing.

3 Name the lines of reference that are used to set out skeleton framed buildings.

4 Define the term *height of collimation*.

5 Explain the purpose of a datum mark and state a method by which it may be established.

6 Describe a method of checking a row of column kicker boxes for line prior to casting.

7 Describe *two* methods used for setting out curves without using the centre point.

Chapter 8

Inspection and safety

It is essential that all concreting and formwork operations are under the control of suitably experienced supervisory staff at all times. The formwork itself must be inspected: during the erection of the formwork; just prior to and during concreting; and finally during the striking.

The purpose of this supervision is to ensure that the work is being carried out in a safe manner; that the formwork meets the engineer's specification in all respects and is capable of producing the required quality of concrete (e.g. dimensional accuracy and specified standard of surface finish); that all box-outs, cast-in fittings and steel reinforcements are according to detail; and finally that the formwork is not struck until the concrete has sufficiently cured to gain the required strength.

The following check procedures are suitable to serve as a basis for site inspections carried out at various times by the formwork supervisor, general foreman, site engineer, clerk of works, or where appointed the falsework/temporary works co-ordinator. Owing to the wide ranging nature of construction works, the list contains items that may not be relevant to a particular job. It should therefore be amended to suit the work in hand.

It is recommended that one of the above suitably qualified persons is given the authority to: issue permission to load or fill the formwork; stop the work in progress should it fail to meet stated criteria; and issue permission to strike the formwork. In addition it is considered essential that a fully experienced formworker is on standby observation duty during the actual pouring of the concrete to deal with any minor adjustments during casting, with replumbing, levelling or lining operations after casting, and any other contingencies.

Before concreting

1. Has the concreting gang got unobstructed access to fill the formwork?
2. Can concrete skips, pump pipelines, conveyors and other mechanical handling devices be brought up to fill the formwork?
3. Are working platforms and ladders required to gain access? If so, are they adequate for the work in hand, safe and securely tied in position?
4. Are all the formwork materials used suitable for the work in hand and free from distortions, defects, rust, rot etc.?
5. Is the formwork in accordance with the design specification; in the right position; of the required dimensions; level, plumb and in line?
6. Are all ties, supports, struts, clamps, braces etc. correctly positioned, tight, and suitably tied or fixed to prevent the possibility of later movement (see later proprietary equipment checklist)?
7. Have all the stop ends, fillets, features, inserts, box-outs, cast-in fittings etc. been accurately located and securely fixed?
8. Has provision been made for ease of striking without causing damage to the concrete, e.g. striking strips in decking?
9. Is there sufficient clearance for the easy removal of both the form ties and the actual forms themselves?

10 Has a release agent been applied to the form face?
11 Have all holes, fixings and joints in the form face been suitably filled or taped to prevent grout leakage?
12 Is the steel reinforcement in accordance with its specification – of the correct size, type, shape, spacing and location?
13 Is the steel reinforcement free from rust scale and release agent contamination?
14 Is the cover to steel reinforcement correct, and are there positive arrangements to maintain such cover?
15 Have the adjoining surfaces of any previously cast concrete been adequately prepared, e.g. by scabbling?
16 Have all forms been cleaned out to remove tie wire loose ends and other debris?
17 Are any steel reinforcement or cast-in fittings to be inserted during concreting operations? If so, are these available on site?
18 Are all items of equipment to hand, e.g. vibrators, screeds, tamps, floats and temporary lighting? If so, are they in safe working order and is a suitable power supply available (if required)?
19 Are curing membranes and insulation materials available, including the means to support them clear of the fresh concrete surface if required?
20 Are helmets, gloves, goggles, ear muffs and other personal protection items available for use by all personnel?

Checking proprietary equipment

Proprietary equipment is extensively used in all formwork construction. A special check to look out for the following common errors should be made during all site inspections:

Column and beam clamps Over-tightening of wedges and bolts, causing crushing of backing timbers and subsequent loss of strength; no support under clamps (nails or blocks) to prevent them from slipping; projecting end of clamps not painted white or bright to make them obvious; incorrect location of or loose wedges.

Floor centres Incorrect location of or failure to tighten locking nuts or wedges; bearing plates inadequately seated; centres used at their minimum span (no provision to retract for striking).

Form ties (Figure 191) Too small or incorrectly positioned washer plates; over-tightening of bolts; unequal size of soldiers or walings; damaged or substandard threads.

Adjustable steel props (Figure 192) Use of visibly bent, damaged or corroded props; use of nails, bolts or steel reinforcement in place of the high tensile steel pin supplied by the manufacturers; missing or inadequate bracing and lacing; props out of plumb (check with a spirit level); props eccentrically loaded; inadequate bearing at either top or bottom.

During concreting

Look out, listen for and order an immediate stop to concreting operations should any of the following occur:

1 Formwork straining noises (creaking timbers, ties snapping etc.)
2 Any opening joints or grout leakage
3 Any movement or distortion in the formwork or bulging of the form face.

Each of these factors may be an indication of a possible formwork blow-out or collapse. Common failures commence with grout leakage followed by distortion and bulging in the formwork. In cases where a form tie fails its neighbouring ties become overloaded and rapid progressive failure can result.

The immediate action is to stop pouring and vibration immediately, evacuate personnel from potentially dangerous situations, and then take remedial action to relieve and control the excessive pressures.

The remedial action taken will vary depending on the particular circumstances applicable to each case. Where minor movement is concerned it may be possible to strengthen and force back into position distorted or bulging formwork by inserting additional form ties, props, clamps, braces etc.

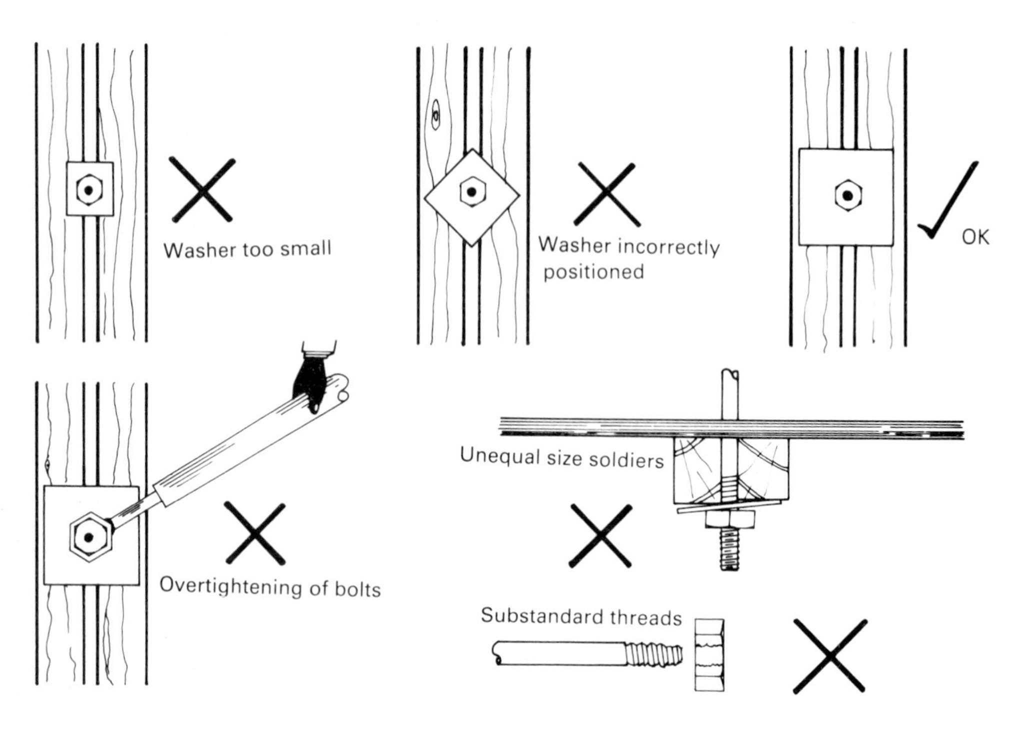

Figure 191 *Form ties*

It can be extremely dangerous to attempt to force formwork back in position where major movement and distortion has taken place. Such action can worsen the situation, and has the potential to cause progressive collapse or overturning of the whole formwork assembly. The correct procedure prior to the recommencement of concreting, where major movement is concerned, is:

1 To remove the fluid concrete from the formwork to a level below that of the distortion or bulge, after which the formwork can be strengthened and straightened; or
2 To remove one form face, clear the concrete and reassemble forms using additional form ties, props etc.; or
3 To wait until the concrete has hardened sufficiently to stay in position, remove one form face, hack off bulging concrete, and reassemble forms using additional form ties, props etc.

Where appropriate, check that all distance pieces have been removed and that cast-in fittings, fixing blocks etc. have been inserted.

On completion of concreting operations, recheck and adjust as required all formwork for plumb, level and line.

Apply curing membranes and insulation as specified.

Clean off any concrete spillage or grout leakage from the outside of the forms before it has a chance to harden.

Remove stop ends to walls and slabs as soon as the concrete has hardened sufficiently to stay in position (the longer they are left in position, the harder they will be to remove).

Before striking

The time taken between the pouring of the concrete and the striking of the formwork will vary according to:

1 The type, size, shape and position of the concrete item, e.g. wall, column, slab or beam
2 The type of concrete used and surface finish required
3 The weather conditions and exposure of the site (mainly temperature considerations)
4 The requirements of the specification.

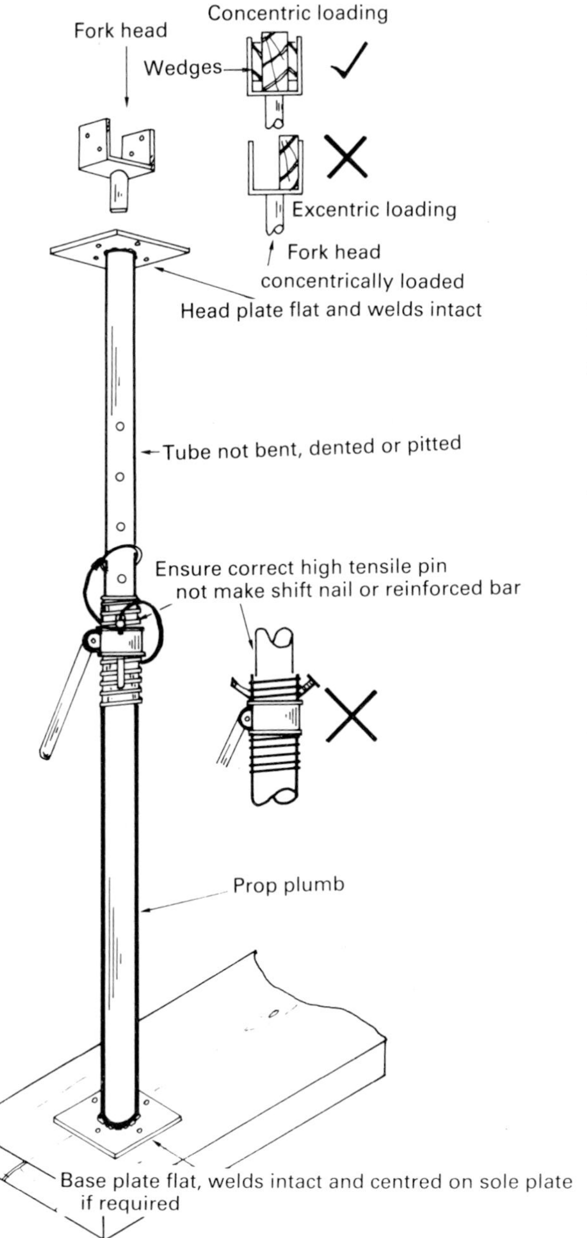

Figure 192 *Adjustable steel props*

Under no circumstances must any formwork be removed until the concrete has cured sufficiently for it to be self-supporting and capable of carrying all loads which will be imposed upon it. Typical striking times are as follows:

Vertical surfaces such as walls, columns and beam sides can normally be struck the day after concreting. In cold weather it may be necessary to delay striking for a further day. In both cases the concrete will still be green and easily damaged, so strike with care. Use hardwood wedges rather than metal nail bars to prise the forms from the concrete.

Suspended horizontal surfaces such as slab and beam soffits normally require formwork support for much longer periods – typically 4 to 6 days, but considerably longer during cold spells. It is normal after striking horizontal formwork to reprop the concrete in strategic positions for a further period.

The final decision to strike must always be taken by the appointed person (site engineer or falsework/temporary works co-ordinator) who has the knowledge and experience of what is involved.

As the safe time to strike formwork is basically dependent on the concrete strength, it can be determined by tests on cubes made from the same mix of concrete and cured under similar conditions.

The information contained in Table 23 can be used as a guide for minimum striking times where ordinary Portland cement or sulphate resisting Portland cement has been used for the concrete mix.

Example
To determine the minimum striking time for slab formwork, assuming a concrete surface temperature or air temperature during curing of 10°C.

Table 23 *Minimum striking times*

Location of formwork	*Temperature* t		
	16°C and over	*7°C*	*Any* t *0–25°C*
Vertical surfaces			
Walls, columns and beam sides	12 hours	18 hours	$300/(t + 10)$ hours
Horizontal surfaces			
Slab soffits	4 days	6 days	$100/(t + 10)$ days
Slab support structure	10 days	15 days	$250/(t + 10)$ days
Beam soffits	10 days	15 days	$250/(t + 10)$ days
Beam support structure	14 days	21 days	$360/(t + 10)$ days

Slab soffit

$$\text{minimum striking time} = \frac{100}{t + 10} \text{ days}$$

$$= \frac{100}{10 + 10} = 5 \text{ days}$$

Slab support structure

$$\text{minimum striking time} = \frac{250}{t + 10} \text{ days}$$

$$= \frac{250}{10 + 10} = 12.5 \text{ days}$$

Striking procedure

The procedure for striking formwork must be carefully considered prior to its commencement. This is in order to avoid any safety risk to personnel; any stress or damage to the structure being built; and, in addition, any stress or damage to the formwork itself. In general striking is the reverse procedure of that adopted when erecting the forms. The following precautions should be observed at all times:

1 All personnel should wear the appropriate protective clothing (safety boots, helmet, gloves etc.)
2 Never allow unauthorized personnel to enter the striking area (temporary barricades and warning signs can be erected).
3 Always release bolts, ties and clamps gradually to prevent binding and damage.
4 Always release props evenly by small increments (half a turn of the handle). Start from the centre prop of each ledger and working outwards towards its ends; then repeat the operation.
5 Always control the striking by removing a section at a time. Crash striking, where large areas of formwork are allowed to fall at once time, should never be permitted. It damages the formwork, is potentially dangerous to personnel, and in addition stresses the structure by its sudden loading.
6 Always clean, denail, repair and store under cover ready for reuse all formwork materials:

   Timber forms Remove all concrete spillage and traces of grout with a stiff brush and a hardwood (not metal) scraper. Coat the form face with a release agent before storage.
   Glass reinforced plastic Wipe over with a wet cloth to remove any traces of grout and dust. Coat with a release agent.
   Metal forms Remove traces of grout with either a wet cloth or a scraper. Coat all surfaces with a light oil or rust inhibitor before storage.
   Ancillary equipment Props, clamps and bolts should be cleaned with a wire brush to remove grout etc., coated with a rust inhibitor, and have their threads lubricated prior to storage.
7 All formwork is best stored in a level horizontal position. Wall panels and column forms should be stacked form face to form face. Bolts, ties and washers should be stored in baskets or bins to prevent losses.

Example
The following is typical procedure to be adopted when striking a section of decking formwork, consisting of plywood soffit, timber joist and ledgers, and supported on adjustable steel props:

1 Remove beam sides and slab edges.
2 Place props centrally under each sheet of plywood (prevents them crashing down).
3 Gradually slacken off and remove props to centre ledger. Use nail bar to prise ledger away from joists and lower to ground.
4 Repeat previous stage on the remaining ledgers in the section being struck, working outwards from the centre.
5 Use nail bar to prise each joist away from the decking in turn and lower to ground.
6 Remove in turn, one at a time, the props supporting the centre of the plywood sheets. Use a hardwood wedge to prise the plywood from the concrete and lower to ground. Replace the prop in position to act as a temporary support for the slab, before removing the next sheet of plywood.
7 Drive hardwood wedge between striking strip and concrete, remove strip and lower to ground.
8 Denail, clean, repair, coat with release agent, lubricate, treat with rust inhibitor and store ready for subsequent reuse all materials and components as applicable.

Use of cranes

Whenever cranes are used on site, including during the erection and striking of formwork and the transportation of concrete, it is essential that all operations are in the total control of an experienced banksman/slinger. This person will have been trained in the general principles of crane operation and slinging (the securing of loads prior to lifting).

S/he must position her or himself so that s/he has a clear view of the crane driver, the lifting position and the load destination. S/he must be readily identifiable to all concerned as the person responsible for such operations.

Communication between the crane driver and the banksman is via a recognized system of hand signals as illustrated by Figure 193, or by radio.

Health and safety

Health and safety requirements for formwork and concrete operations do not generally differ from other areas of construction work.

The Health and Safety at Work Act 1974 (HASAWA) is the main statutory piece of legislation that covers the health and safety of all persons at their place of work and protects other people from risks occurring through work activities. In addition earlier health and safety requirements run parallel to HASAWA; the Factories Act 1961 and the four parts of the Construction Regulations are the main examples.

HASAWA is an enabling Act, which means that an order relating to any safety topic simply has to be laid before Parliament in order to become law. By this means it is intended to gradually replace all earlier safety legislation with new regulations and codes of practice made under HASAWA, e.g. Safety Signs Regulations 1981, Health and Safety (First Aid) Regulations 1981, etc.

HASAWA places responsibilities for health and safety on all parties in the construction industry. This means employers and management; employees; self-employed people; designers, manufacturers and suppliers of equipment and materials.

Employers' and management duties
Employers have a general duty to ensure the health and safety of their employees, visitors and the general public. This means that the employer must:

1 Provide and maintain a safety working environment
2 Ensure safe access to and from the workplace
3 Provide and maintain safe machinery, equipment and methods of work
4 Ensure the safe handling, transport and storage of all machinery, equipment and materials

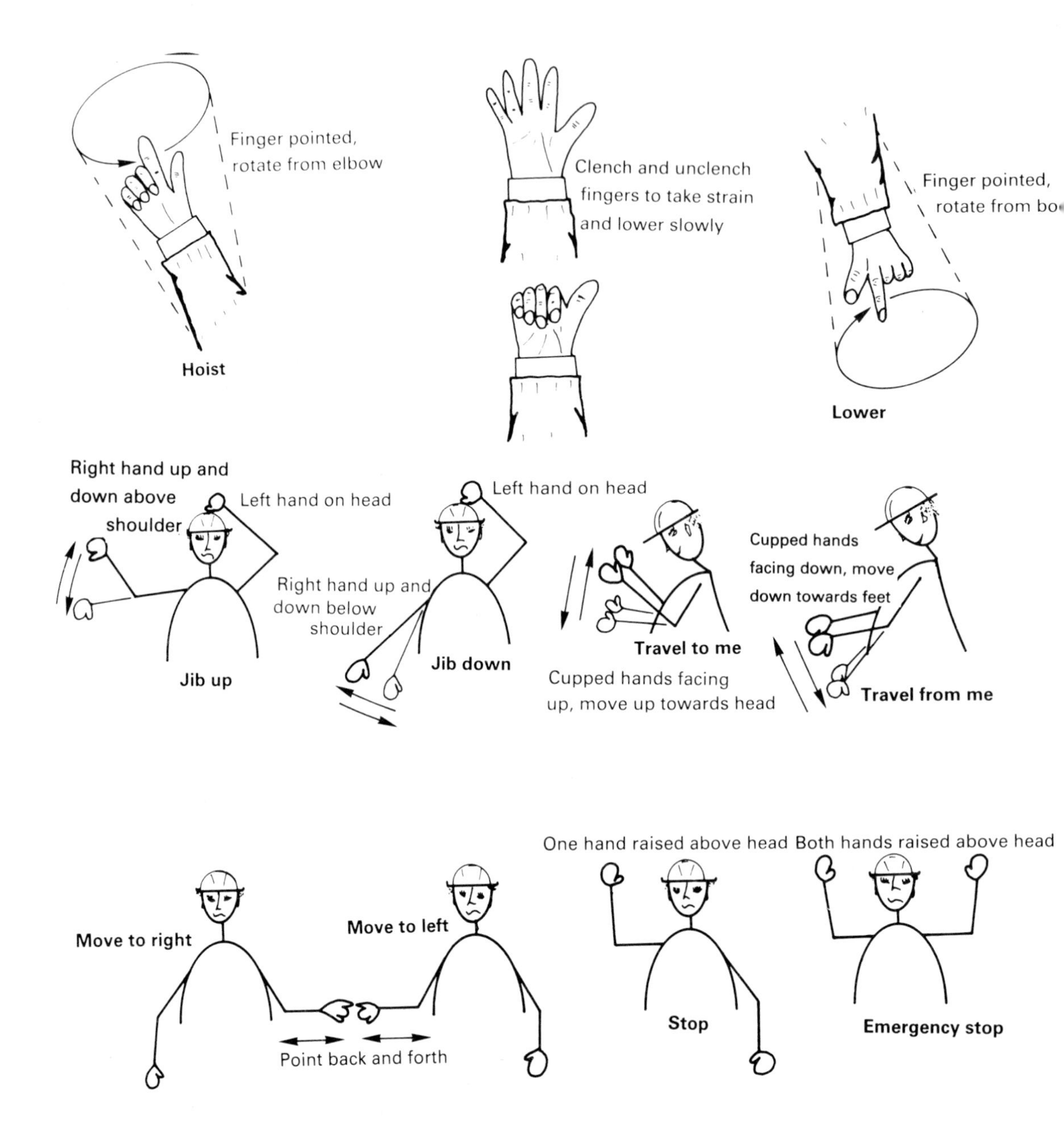

Figure 193 *Banksman handsignals to crane driver*

5 Provide their employees with the necessary information, instruction, training and supervision to ensure safe working
6 Prepare, issue to employees and update as required a written statement of the firm's safety policy
7 Involve trade union safety representatives (where appointed) with all matters concerning the development, promotion and maintenance of health and safety requirements
8 An employer is not allowed to charge an employee for anything done or equipment provided to comply with any health and safety requirements.

Employees' duties
An employee is an individual who offers his or her skill and experience etc. to his or her employer in return for a monetary payment. It is the duty of every employee while at work to:

1 Take care at all times to ensure that his or her actions do not put at risk himself or herself, workmates or any other person
2 Co-operate with his or her employer to enable them to fulfil the employer's health and safety duties
3 Use the equipment and safeguards provided by the employer
4 Never misuse or interfere with anything provided for health and safety.

Self-employed duties
The self-employed person can be thought of as both his or her employer and employee. Therefore the duties under the Act are a combination of those of the employer and employee.

Designers', manufacturers' and suppliers' duties
Under the Act, designers, manufacturers and suppliers as well as importers and hirers of equipment, machinery and materials for use at work have a duty to:

1 Ensure that the equipment, machinery or material is designed, manufactured and tested so that when it is used correctly no hazard to health and safety is created
2 Provide information or operating instructions as to the correct use, without risk, of their equipment, machinery or material (employers should ensure that this information is passed on to their employees)
3 Carry out research so that any risk to health and safety is eliminated or minimized as far as possible.

Enforcement of HASAWA
A system of control was established under HASAWA. This is aimed at reducing death, injury and ill health due to work related activities.

The Health and Safety Executive (HSE) is divided into a number of specialist sections or inspectorates, each dealing with a specific area of work. The section with the main responsibility for the construction industry is the Factory Inspectorate. Factory inspectors are based in local offices situated throughout the country, from which visits are made to individual workplaces and from whom information can be sought.

Factory inspectors have the power of entry, examination and investigation in order to assist them in the enforcement of the HASAWA and the earlier safety legislation. In addition to giving employers advice and information with regard to health and safety enquiries, an inspector can:

Enter premises in order to carry out investigations, including the taking of measurements, photographs, recordings and samples. The inspector may require the premises to be left undisturbed while the investigations are taking place.
Take statements An inspector can ask anyone questions relevant to the investigation and also require them to sign a declaration as to the truth of the answers.
Check records All books, records and documents required by legislation must be made available for inspection and copying.
Give information An inspector has a duty to give employees or their safety representative information about the safety of their workplace

and details of any action they propose to take. This information must also be given to the employer.
Demand The inspector can demand the seizure, dismantling, neutralizing or destruction of any machinery, equipment, material or substance that is likely to cause immediate serious personal injury.
Issue an improvement notice This requires the responsible person (employer, manufacturer etc.) to put right within a specified period any minor hazard or infringement of legislation.
Issue a prohibition notice This requires the responsible person immediately to stop any activities which are likely to result in serious personal injury. This ban on activities continues until the situation is corrected. An appeal against an improvement or prohibition notice may be made to an industrial tribunal.
Prosecute All persons, including employers, employees, the self-employed, designers, manufacturers and suppliers, who fail to comply with their safety duty, may be prosecuted in a magistrates' court or in certain circumstances in the higher court system. Conviction can lead to unlimited fines, or a prison sentence, or both.

The Construction Regulations

These are the regulations made under the Factories Act 1961 which are specific to construction operations. They are divided into four parts, each dealing with a different aspect of work:

1 Construction (General Provisions) Regulations 1961 set out minimum standards to promote a good level of general safety.
2 Construction (Lifting Operations) Regulations 1961 lay down requirements regarding the manufacture, maintenance and inspection of lifting appliances used on site (gin wheels, cranes, hoists etc.).
3 Construction (Health and Welfare) Regulations 1966 set out minimum provisions for site accommodation, washing facilities, sanitary conveniences and protective clothing.
4 Construction (Working Places) Regulations 1966 control the erection, use and inspection of scaffolds and other similar temporary structures.

General safety

It should be the aim of everyone involved in the construction industry to prevent accidents. Remember that you are required by law to be aware of and fulfil your duties under the Health and Safety at Work Act.

The following are some of the basic measures which are required to ensure safety for typical formwork and concrete operations (Figure 194).

The most common cause of accidents is where people fall or are themselves struck by falling objects. Therefore ensure the following:

1 Safety helmets are worn.
2 Excavations are protected by barriers.
3 Inspection chambers in the ground and openings in a slab such as service ducts, lift shafts and stair wells are either covered over and a warning sign erected or protected by some form of guard rails.
4 Scaffolds and other working platforms, or the decking edge where no separate working platform is provided, should be equipped with a guard rail, toe board and wire guards.
5 Where guarding to working area is impractical the use of safety nets and harnesses must be employed. Safety nets also have the added advantage of catching any accidentally dropped tools, plant or materials, thus preventing injury.
6 Tidy working areas are safer than untidy ones. Remove periodically all unnecessary obstructions which may create a hazard, e.g. offcuts of material, unwanted materials, disused items of plant etc.
7 Waste materials being disposed of from heights should always be lowered safely (crane, hoist, shute etc.) and never thrown or dropped from the working area.
8 No person should be allowed access under slab formwork during pouring, as a few makeshift props, pins, inadequate bracing or uneven loading can set off a chain of events leading to progressive collapse.
9 Erect warning signs during striking operations.
10 Never allow partially struck formwork to be

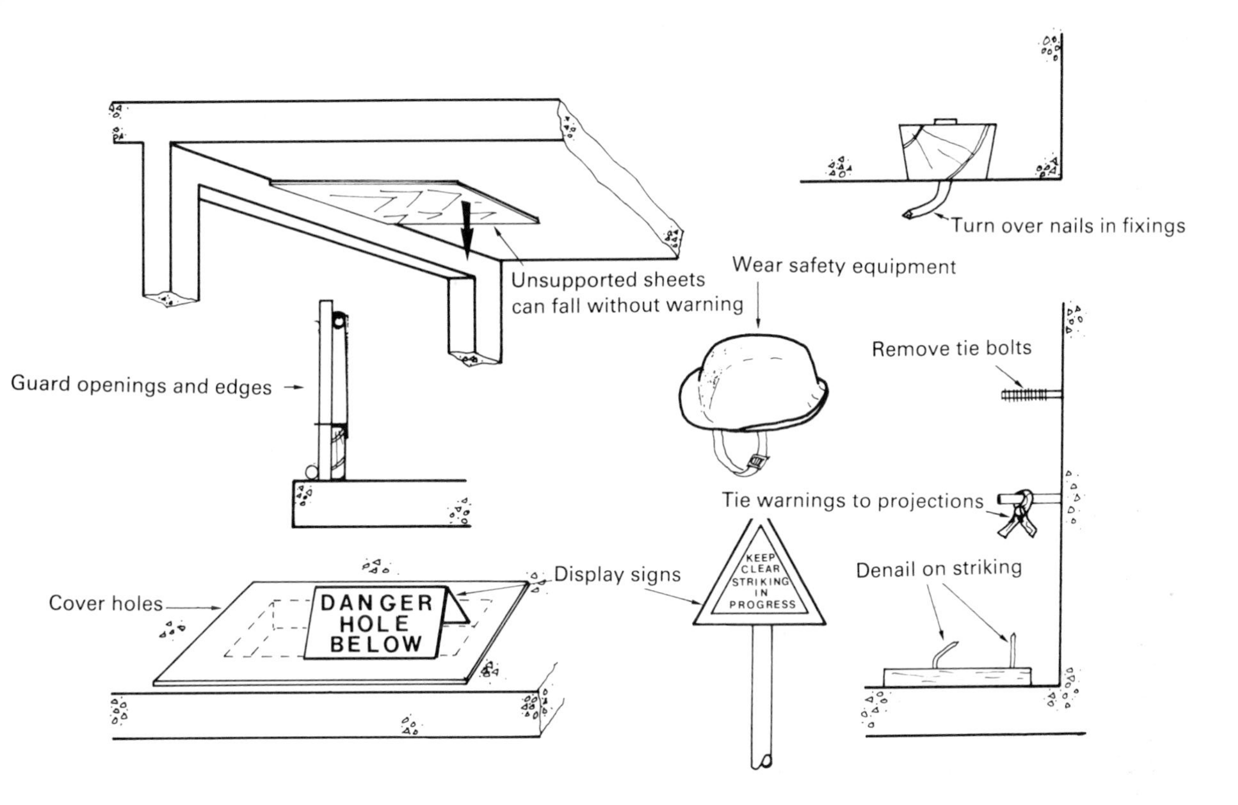

Figure 194 *General safety measures*

left unattended. A sheet of plywood decking stuck to the concrete after its supports have been removed can fall without warning.

Other precautions are as follows:

1 Ensure for your own safety that you wear the appropriate protective clothing. In addition to a safety helmet, safety footwear (steel lined toecap and soles) should also be worn at all times. Use ear protectors when carrying out noisy activities. Wear safety goggles when involved with any operation that is likely to produce dust, chips or sparks; dust masks or respirators where dust or fumes are present; and gloves when handling material. Finally, wear wet weather clothing for inclement conditions. With the exception of safety footwear, these items should be supplied by your employer free of charge.
2 Safety signs and barriers should be erected to keep unauthorized persons away from areas when form erection, concreting or striking is in progress.
3 Denail all formwork on striking to prevent stabbing injuries.
4 Turn over nails and other fixings that project from the concrete.
5 Projecting starter bars, bolts etc. should be made obtrusive by tying warning tape to them.
6 Electric leads, compressed air lines and pump lines must be kept as short as possible and routed safely out of the way, to prevent the risk of tripping injuries or damage due to plant running over them.
7 All scaffolds, working platforms, excavations, cranes and other lifting and handling equipment, such as gin wheels, chains, slings and hooks etc., are subject to periodic statutory inspections, normally by the site safety

officer (see Construction Regulations). A signed record of all inspections must be maintained on site and made available to the factory inspector if required.

8 In general no person under the age of 18 years should be allowed to operate any item of powered plant, e.g. wood cutting machine, abrasive wheel, crane, power float etc. In all situations persons must be instructed as to the safe operating procedures of a particular item of plant before use. This instruction must be carried out by a person who are themselves fully competent in the use of the item.

Finally, take care of your personal hygiene. This is just as important as any physical protection measures. Certain building materials, e.g. cement, admixtures and release agents, can have an irritant effect on skin contact; are poisonous if swallowed; and can result in narcosis if their vapour or powder is inhaled. By taking proper precautions these harmful effects can be avoided. Follow manufacturers' instructions; avoid inhaling spray mists, fumes and powders; wear disposable gloves or a barrier cream; and thoroughly wash hands before eating, drinking, smoking and after work.

Self-assessment questions

Question	*Your answer*
1 List *ten* points that should be included in the check of wall formwork before concreting.	
2 Explain the difference between the Health and Safety at Work Act and the Construction Regulations.	
3 Briefly describe the duties of the formworker on standby duty during concreting operations.	
4 State the signal that a banksman would give to a crane driver when an emergency stop is required.	

5 List *four* points of safety concerning the inspection and use of adjustable steel props.

6 Describe the factors that affect formwork striking times.

7 Which one of the following personnel is responsible for authorizing formwork striking?
(a) Ganger man
(b) Health and safety inspector
(c) Formwork carpenter
(d) Falsework co-ordinator.

a [ ]	b [ ]	c [ ]	d [ ]

8 Briefly describe the maintenance procedures after a striking for each of the following formwork materials:
(a) Timber form faces
(b) Metal form faces
(c) Glass reinforced plastic form faces
(d) Ancillary metal equipment.

Chapter 9

Glossary of terms

Adjustable steel prop (ASP) A proprietary prop with provision for adjustment so that its length may be varied. Used either as vertical supports for slab formwork or as inclined members to hold vertical formwork plumb.
Admixture A material added to the basic constituents of a mix to alter one or more of its properties.
Aggregate A filler material in concrete.
Aggregate/cement ratio The relationship between the amounts of aggregate and cement in a concrete mix.
All-in aggregate A mixture of fine and coarse aggregates.
Batching The process of proportioning the constituent materials for a concrete mix, either by weight or by volume.
Beam box The formwork for the sides and soffit of a beam.
Beam clamp A timber yoke or proprietary clamp which helps to maintain a constant beam width during casting.
Bleeding Excess water rising to the surface of freshly placed concrete. This leaves behind a network of interconnected voids which reduces both the strength and the durability of the concrete.
Blow hole A small hole or cavity in the concrete face due to air pockets trapped against the form face. Mainly associated with impervious form faces; or excessive application of release agents; or use of neat oil as release agent.
Box-out The formwork which forms an opening or pocket in a concrete item, e.g. service duct opening, window or door opening, bolt pocket.
Bracing A diagonal member in formwork used to restrain sideways movement. Depending on the loading conditions, may act either in tension or in compression.
Bulk density Mass per unit volume of a material including its air voids.
Bulking The increase in damp aggregate volume due to the film of liquid separating the particles.
Cement A fine powder which, when mixed with water, forms a paste that gradually hardens. Acts as an adhesive in concrete bonding aggregate together.
Coarse aggregate Having particles mainly greater than 5 mm in size.
Cohesion The attraction between the constituents of concrete causing them to stick together.
Column box The assembled formwork for a column.
Column clamp A timber yoke or proprietary steel clamp which holds the sides of column boxes tightly together during casting. These are positioned closer together towards the bottom of the form owing to the increase in hydrostatic pressure.
Compacting factor test A test used to assess the workability of fairly dry mixes. It compares the mass of concrete partially compacted by a fixed amount of work to that of fully compacted concrete.
Compaction The process of removing trapped air in freshly placed concrete.
Concrete A composite structural material, made mainly from a mixture of cement, aggregate and water.

Construction joint A joint formed to allow a break in concreting operations. Subsequently fresh concrete is placed against the joint, which when hardened will be expected to provide structural continuity.
Contraction joint A joint formed in concrete to allow the natural contraction on drying and that due to tensile stress.
Cover The amount of concrete surrounding steel reinforcement. Provides resistance to corrosion and fire.
Cracking Also termed plastic cracking. Its minor form is known as crazing. A defect where the concrete surface is broken up by a network of cracks of varying widths caused either by cooling and drying out (plastic) or by tensile stress under load.
Cube test A test carried out on hardened concrete cubes to determine their crushing strength in N/mm^2, normally at 28 days after casting.
Curing The process of retaining water in recently poured concrete and providing insulation from extremes of temperature.
Curtains Hardened grout runs down the face of completed work. Often accompanied by grout loss or honeycombing caused by poor sealing between formwork and completed structure.
Datum mark Positions indicated on walls and columns to provide a reference point for positioning other building elements and components. Also see *Ordnance datum*.
Day joint See *Construction joint*.
Dead load The load carried by formwork, consisting of the weight of the concrete, its steel reinforcement and the self-weight of the formwork.
Decking The horizontal form face to slab soffits.
Density Mass (kg) per unit volume (m^3) of a material.
Distance piece A short member used to ensure the correct spacing of parallel wall and beam forms.
Double headed nail A two headed wire nail, also known as a formwork nail. The first head is driven home for strength, leaving the second projecting to enable easy withdrawal.
Draw The lead or taper of a fillet or box-out which permits easy removal from the cast concrete. A false lead or draw may be provided for square or large features.
Durability The ability of concrete to resist the effects of weather, chemical attack and abrasion.
Edge form The vertical formwork to the edge of a slab, path or road.
Environmental load The load on formwork caused by wind action, accidental overloads etc.
Expanded metal A sheet metal with an open expanded mesh which can be used for forming box-outs and permanent stop ends. When used as stop ends it has the advantage of providing a good key for the subsequent pour.
Expanded polystyrene An air expanded plastic obtainable in blocks and sheets. Mainly used for forming small box-outs and pockets for holding-down bolts, balusters, handrails etc. The polystyrene is removed after casting either by breaking it up, setting it alight and burning out, or by softening it out with petroleum spirit.
Fair face A plain concrete finish of high quality produced straight from the form, without any subsequent touching up or making good.
Falsework The temporary support structure which supports the form face. A term mainly used to describe the support structure for major civil engineering works such as bridges.
Field settling test An on-site test used to determine the percentage of silt present in a fine aggregate.
Fillet A small section of material which is fixed to the form face to produce a feature, groove, bullnose, chamfer etc.
Fine aggregate Having particles mainly less than 5 mm in size.
Floor centre A proprietary metal beam of either box or lattice design. These are adjustable in length and are used in place of ledgers in slab formwork.
Form face The form surface which will be in contact with the cast concrete.
Form filler A filling compound used for repairs and filling small gaps, nail and screw holes for fair faced work. Repaired areas should be overfilled and sanded flush after the filler has hardened.

Form lining The material used to line the inside of a form in order to produce the desired concrete finish. It is thus the form face.
Form panel A standard framed panel consisting of sheathing and supporting stiffening members. These can be either proprietary with ply or metal sheathing, or made up from timber framework and ply sheathing. Both are normally based on standard sheet sizes and are used for a wide range of work, e.g. wall forms, slab construction and base boards for precast work.
Form tape A waterproof self-adhesive tape which is used to seal joints in the form face to prevent grout leakage.
Form ties Steel bolts or tie rods, also known as wall ties, used to anchor the vertical retaining sides of formwork together.
Gang mould A series of moulds for precasting more than one concrete unit at a time.
Grading The proportion of different particle sizes in a sample of aggregate.
Grout loss The leakage of cement and water at form joints, bolt positions etc. Causes a surface defect having a sand textured appearance lacking in cement paste.
Honeycombing A surface defect having a coarse stony surface with air voids, lacking in cement paste and fine aggregate. Can be caused through excessive leakage at formwork joints, a poor placing method, mix segregation or inadequate compaction.
Hydrostatic pressure The pressure exerted within formwork by the fluid concrete.
Imposed load The load placed on formwork by construction workers, concreting equipment and materials etc.
Insert A dovetailed timber block or proprietary item that is cast into the concrete and provides a subsequent fixing point.
In situ cast concrete A concrete unit that has been cast in the actual position or location where it will be used.
Joist A timber or proprietary aluminium member which is used to support the decking material of floor slabs.
Key A small indent in a concrete surface or a tooling of a concrete surface to remove the fine material and expose the aggregate, thus ensuring a good bond.
Kicker A small concrete upstand cast above a slab. The plan size is the same as the finished wall or column but only about 150 mm in height. Its purpose is to locate the bottom of the subsequent formwork and to prevent grout loss.
Lacing Horizontal members used in conjunction with adjustable steel props to space and tie them together. Their use greatly increases the safe working load.
Lean concrete A mix having a high aggregate/cement ratio.
Ledger A timber or proprietary aluminium member which is used to support the joists of slab formwork. These are also known as runners or primary beams.
Lift The vertical height of concrete that is to be poured in one operation.
Live load See *imposed load.*
Mismatch/misalignment A step, wave or other deviation in the concrete surface. Caused by inadequate formwork detail; loose ties, clamps and props; or concrete discharge from a height with a too rapid rate of pour.
Mould The formwork for a precast concrete item.
Movement joint A joint formed in concrete to allow expansion and contraction from the original position.
Ordnance datum The mean sea level at Newlyn in Cornwall to which Ordnance bench marks and subsequently on-site temporary bench marks are referred.
Permanent form The formwork or form face that is to be left permanently in position after the concrete has been cast.
Placing As *pouring.*
Pouring The process of placing concrete into formwork.
Precast concrete A concrete item that is cast out of its final location, either on site or in a factory.
Quartering The reduction of a material sample into four equal parts, two of which are remixed in order to obtain a representative sample.
Ready mixed concrete Batched and mixed by a ready mix specialist and transported to site in mixer trucks.

Reinforcement Steel embedded into concrete to resist tensile and shear stresses.
Relative density The relationship between a material density and the density of water.
Release agent A substance that is applied to formwork in order to prevent adhesion between a concrete surface and the form face.
Ribbon A horizontal member fixed to soffits which prevent beam sides and slab edges spreading. They may also be used to strut off.
Rich concrete A mix having a low aggregate/cement ratio.
Sheathing The form face of vertical forms.
Sieve analysis A test to determine the grading of an aggregate.
Slump test The most common on-site test for workability. Involves measuring the distance that a standard cone of concrete collapses when its support is removed. Useful for comparing the consistency between different mixes.
Snots Hardened grout runs/fins hanging from the underside of soffits. The removal of snots prior to soffit finishing is termed 'snotting'. Caused through the use of damaged or poorly jointed decking.
Soffit The underside of a floor slab or beam. Soffit formwork is called decking.
Soldier A vertical member used to stiffen and anchor vertical forms.
Soleplate A horizontal member, also known as a sill, which is placed under vertical supports to ensure even distribution of the load.
Spalling A surface defect where areas of concrete are misplaced from the hardened surface. Can be the result of damage due to poor striking, or of either frost action or corrosion of reinforcement.
Stop end A member used to form a day or construction joint at the end of a day's concreting.
Stress A concrete member subjected to a force, in either compression, tension or shear.
Strike To strike or striking is the process of removing the formwork from the face of the cast concrete. This is carried out after the concrete has cured sufficiently to become self-supporting.
Strongback A proprietary metal soldier for use where high pressures are anticipated.
Structural grid The intersecting centre lines of the columns, walls and beams in skeleton framed buildings.
Strut A compression member designed to hold vertical formwork in position.
Table form A proprietary decking and support system for soffits which can be struck, crane handled and repositioned in one unit.
Tie A tension member designed to maintain parallel distances between forms. See *distance piece* and *form ties*.
Trough form A hollow rectangular shaped form used to make trough shaped recesses in the concrete soffit. Normally made of glass reinforced plastic.
V-B consistiometer test A workability test more suited to a laboratory than on site. It measures the time in seconds (V-B degrees) taken to compact a standard cone full of concrete.
Vibration The compaction of freshly placed concrete by mechanical means.
Vibrator A mechanical device to impart vibrations in the concrete which aids its compaction. Can be either the immersion type, known as pokers, or an external vibrator which is fixed to the outside of forms.
Voids Air spaces in concrete formed between aggregate particles.
Waffle form A hollow square-on-plan form used to make recesses in the concrete soffit. Normally made of glass reinforced plastic.
Waling A horizontal member used to stiffen and anchor forms, often in conjunction with soldiers or strongbacks and form ties.
Water/cement ratio The relationship between the amounts of water and cement in a concrete mix.
Workability The ease with which a concrete mix can be placed and compacted.
Yoke An arrangement of members which encircle beam or column forms to secure them and prevent movement. Proprietary beam and column clamps have in the main superseded made-up yokes.

Self-assessment questions

Question	*Your answer*
1 A rich concrete is one having a: (a) High water/cement ratio (b) High aggregate/cement ratio (c) Low water/cement ratio (d) Low aggregate/cement ratio.	a [ ] b [ ] c [ ] d [ ]
2 Explain what is meant by concrete bleeding.	
3 Distinguish between grout loss and honeycombing.	
4 Distinguish between waffle and trough forms.	
5 Define and state an example of imposed and dead formwork loadings.	
6 Describe the following formwork items: (a) Kicker (b) Ledger (c) Table form.	

7 Plywood used for a vertical form face is termed:
(a) Decking
(b) Soffit
(c) Sheathing
(d) Lagging.

a [ ]	b [ ]	c [ ]	d [ ]

Examination and study guide

Examinations are intended to measure your achievement of the course aims and objectives. In setting examination questions, the examiner is giving you the opportunity not only to demonstrate your knowledge and understanding of a particular topic area but also to apply that knowledge and understanding in order to produce practical solutions for given situations. The aim of the examiner is therefore to obtain a true assessment of your abilities and not, as many think, to mislead, confuse or fail you.

Types of examination

Two distinct types of examination are used when assessing student achievement in technical subjects:

Multiple choice examinations are normally used for the early examinations in a subject area. These papers typically contain 60 questions to be completed in 2 hours. Four possible answers are given, of which only one is correct for each question. The candidate is required to choose the correct answers to each question.
Written examinations are in the main used for the advanced or higher level examinations in a particular subject. The candidate may be required to give a written account, carry out a calculation, produce a sketch or scale drawing, or carry out a combination of these. The number of questions to be answered and the time allowed for these papers vary widely depending on the type of question and the level of the paper. Typical examples are: ten questions in 2 hours; choose six from eight questions in 3 hours; formwork and concrete construction examination, 4 hours.

Multiple choice questions

Multiple choice examination papers are compiled by choosing questions from an item bank in accordance with an examination specification. Both the specification and the items in the bank would have been produced by subject specialists drawn from education and industry. Prior to their use in a live examination, all of the items are edited, pretested and analysed. The most popular type of multiple choice question (and that almost always used by the City and Guilds of London Institute) is known as a four option question. This consists of a question or statement called a *stem* followed by four different answers called *options*. The correct option is called the *key* and the other three options are known as *distractors*. The distractors are such that they will seem plausible to the person with limited knowledge, but will be clearly wrong to the more knowledgeable.

Example
Stem
Which of the following types of form face material has the best reuse potential?

Options
(Distractor) (a) Glass reinforced plastic
(Distractor) (b) Film overlaid plywood
(Distractor) (c) Smooth timber boards
(Key) (d) Steel.

Answer
As 'steel' is the key, the correct response is (d).

The stem of a multiple choice question may refer to a figure, as in the following example.

Example
The result of the slump test shown in Figure 195 is called:

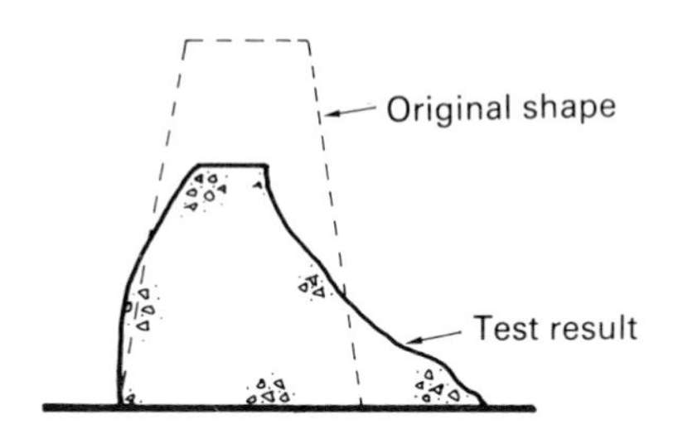

Figure 195 *Slump test result*

(a) True
(b) Shear
(c) Stressed
(d) Collapsed.

Answer
(b)

Occasionally variants on the four option multiple choice question are used in examinations, as in the following example.

Example
Match the items in list one with the items in list two.

List one
Refer to Figure 196.

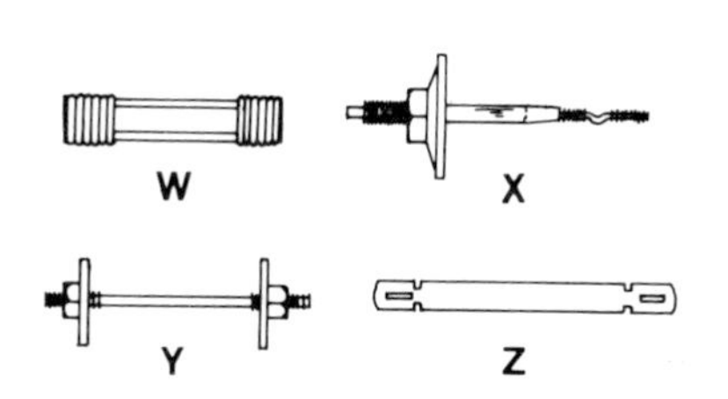

Figure 196

List two
1 Coil tie
2 Loop tie
3 Snap tie
4 She bolt
5 Through tie.

The correct match is:

	W	X	Y	Z
(a)	5	3	4	2
(b)	1	2	5	3
(c)	3	5	2	4
(d)	1	4	5	3

Answer
This question requires you to work through the lists matching the items up (it is usual for the lists to be of different lengths. In this case,

W is 1
X is 4
Y is 5
Z is 3

Therefore the correct response is (d).

Example
The following operations have to be carried out to erect a kicker box:

1 Assemble form sides around kicker.
2 Fix column clamps.
3 Plumb up.
4 Cast kicker.
5 Apply release agent.

The correct sequence for carrying out these operations is:

(a) 1 5 2 4 3
(b) 4 5 1 2 3
(c) 4 1 5 2 3
(d) 4 5 1 3 2

Answer
The correct sequence is:

4 Cast kicker.
5 Apply release agent.
1 Assemble form sides around kicker.
2 Fix column clamps.
3 Plumb up.

Therefore the response should be (b).

Recording your response

The majority of multiple choice examinations are marked by computer. Your responses are recorded on a preprinted answer sheet by filling in the line below the key to each question, e.g.

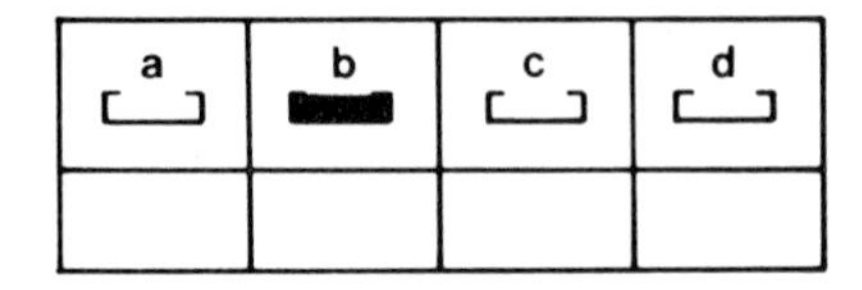

This shows the key to be (b).

After completing all the questions in an examination you are advised to go back through the paper and recheck your work. If you have a change of mind on a question the key can be changed by filling in the large box under your first response (this deletes it) and then filling in the line under your new response, e.g.

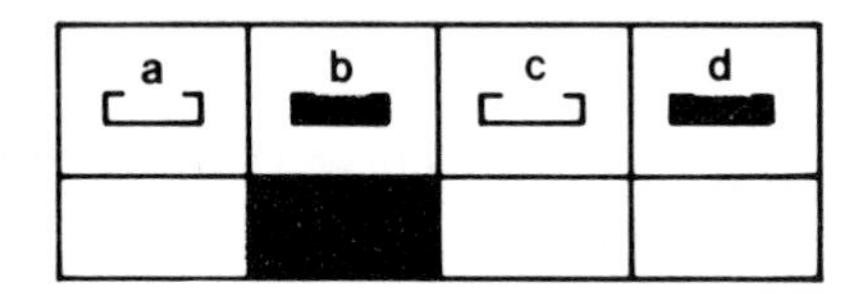

This changes the key from (b) to (d).

Written examinations

Subject specialists from education or industry set draft question papers, based on an examination specification. These draft papers are then submitted to a meeting of the subject moderating committee, who are also subject specialists. At this meeting the draft paper is discussed as a whole and each question considered individually in detail. As a result of discussion between the examiner and moderators, questions may be replaced or amended. This process should ensure that the final version of the examination conforms to the specification and is technically correct, current practice, clear, straightforward and valid.

There are three main types of written answer questions that are used in technical examinations:

1 Short answer questions
2 Structured questions
3 Extended answer questions.

Short answer questions

These consist of one or more problems for which a limited written answer is required. This may vary in length depending on the topic, from one or two words to several sentences. Alternatively a sketch or simple calculation may be required.

Example

State the meaning of the following abbreviations:

(a) OPC
(b) GRP
(c) SC4
(d) MSS
(e) G1S

Typical answer

(a) Ordinary Portland cement
(b) Glass fibre reinforced plastic
(c) Strength class 4; a class 4 stress graded timber
(d) Machine special structural; a machine stress graded timber
(e) Good one side; a plywood sheet with one good face.

Structured questions

These normally commence with a statement containing a certain amount of relevant information which is designed to set the scene for the question. This is then followed by a series of related subquestions in logical order. The length of the expected answer to each subpart will again vary, depending on the topic and the wording of the question, from one or two words to a paragraph, a scale drawing, a detailed sketch, a calculation, or a combination of these.

At each stage the wording of the question will make it clear what is required. Look for words that indicate the expected length or precision required: 'Name', 'list', 'state', 'explain', 'describe', 'find', 'calculate', 'sketch', 'draw' etc. Questions or subquestions that use the words 'name', 'list', 'suggest' or 'state' normally require a brief answer from one or two words up to a sentence at the most. Words such as 'explain', 'define' or 'describe' require a longer answer, although the expected answer may be shortened by the inclusion of the word 'briefly, e.g. 'briefly explain' etc. 'Sketch' or 'find' require less precise or accurate answers than do 'develop', 'draw' or 'calculate'.

Example
Details of a mushroom head that is to be formed at the top of a previously cast square column are shown in Figure 197.

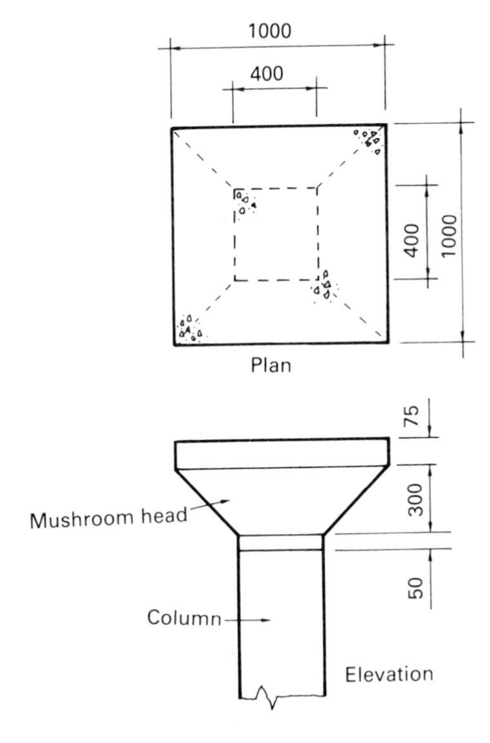

Figure 197 *Mushroom head column*

(a) Sketch a suitable timber and plywood formwork arrangement for casting the mushroom head.
(b) Suggest suitable grades for the timber and plywood.
(c) Name a suitable release agent, and state its method of application.
(d) State briefly in what circumstances would the use of a steel or glass fibre reinforced plastic form be justified.

Typical answer

(a) Figure 198.
(b) Timber: PAR SC4 softwood, plywood Douglas fir G1S.
(c) Chemical release agent applied by spray.
(d) The use of a steel or GRP form can be justified on economic grounds where a large number of mushroom heads have to be cast.

Extended answer questions
These are the more traditional type of question which require an extended, unguided answer to a particular problem. This may be an essay, a scale drawing or a geometrical development. This type of question, particularly the essay version, is rarely set for the technical examinations being considered. The preference is for either the short answer or the structured questions.

Example
Discuss hydrostatic pressure and its development within formwork.

Approach to answer
Clearly an analysis of the question must be made and an outline structure determined for the question. This may consist of:

1 Definition of hydrostatic pressure
2 Factors affecting hydrostatic pressure:
   Density of concrete
   Workability
   Rate of placing
   Method of discharge
   Vibration
   Height of pour
   Dimensions of section
   Reinforcement details
3 Design hydrostatic pressure
   Height of pour
   Arching limit
   Stiffening of concrete
4 Conclusion.

Using this structure as a guide the information on each area can be added, forming a clear logically developed answer.

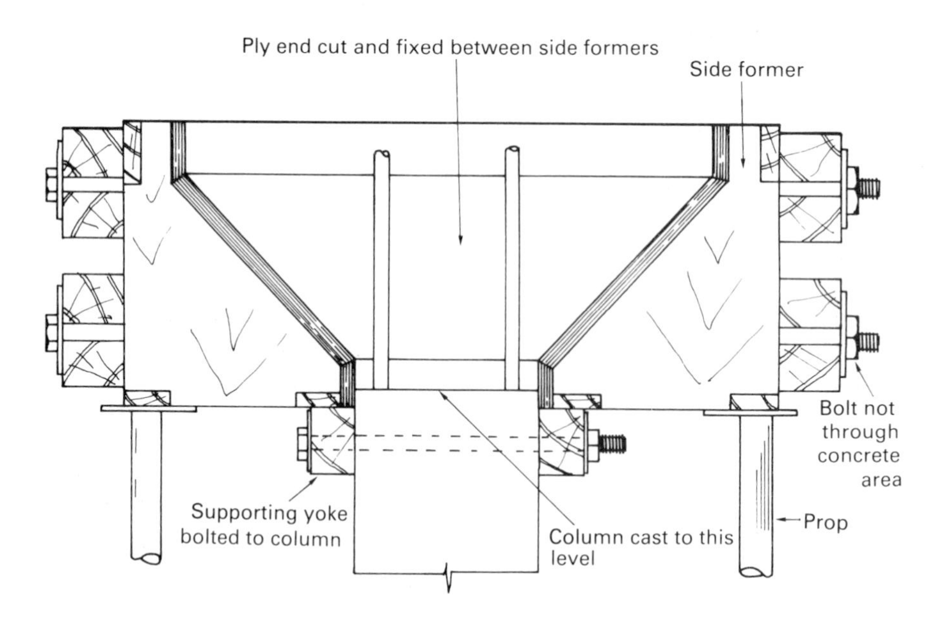

Figure 198 *Plywood mushroom head form*

Marking of written examinations

The marking of completed examination scripts is carried out by a team of examiners under the supervision of a chief examiner. At the moderating meeting where the draft paper was discussed an outline solution and marking scheme for each question will have been agreed. The following example shows a typical question, outline solution and marking scheme.

Example

Define briefly *each* of the following form work terms:

(a) Gang mould
(b) Ledger
(c) Waling
(d) Soldier
(e) Kicker.

	Possible marks
(a) Gang mould: Precast mould to cast more than one unit at a time	2
(b) Ledger: Horizontal timber to carry joists	2
(c) Waling: Horizontal member to stiffen forms	2
(d) Soldier: Vertical member to stiffen forms	2
(e) Kicker: Concrete upstand to locate form and prevent loss of grout	2
Maximum total	10

The team of examiners will meet before marking commences to decide on a common approach to marking. Each examiner will have to forward to the chief examiner samples of their marking. From these samples the chief examiner can tell if any particular examiner is marking too strictly or too easily and advise them accordingly.

You should remember that:

1 Marks can only be given for correct information that is relevant to the question.
2 Marks will not be taken away for the wrong information.

3 Marks are not deducted for ungrammatical expressions or poor spelling.
4 Marks are awarded for each question separately. Care is taken not to let a poor performance or a good performance in one answer reflect in the marks for the rest of your paper. Each answer will get the marks it deserves.
5 A few days after the examination an examiner will receive several batches of papers; one of these will be yours. An impression comes across immediately. Is the paper neatly laid out, and are the sketches well proportioned, or is the paper untidy and are the sketches scrappy? One thing is certain: anything you can do to make the examiner's job easier will be appreciated.

Examination preparation

Examination papers are not set with the intention of tricking you. It is not the examiner's fault if you fail to understand or misread a question and give an answer that has not been asked for. The main causes of student failure are:

1 Lack of preparation
2 Lack of topic knowledge or understanding
3 Bad examination technique.

The first two of these causes can be overcome by following a sensible revision programme.

Revision

Revision is an extremely important part of study. Unless facts and information are repeated, used or revised, they will soon be forgotten. Therefore ideally revision should be a continual process, starting at the very beginning of your course and not just at the end of it, although naturally it will become more concentrated as examination time approaches.

There is no one ideal method of revision, as a method that is successful for one person may not be so for another. Therefore as a guide some of the following techniques could be included in your own study programme:

1 Rough class notes should be rewritten after each lesson.
2 The underlining of key points in textbooks, class handouts and notes may help, as remembering these words can bring back to mind the complete topic.
3 Write brief revision notes. These condense a lot of information into a skeleton of a topic that can be recalled at a later date.
4 Many people find that producing a note network diagram is helpful and can easily be recalled in an examination as an outline to build on. For example, a note network diagram showing the factors which affect the development of hydrostatic pressure within formwork might look like Figure 199.
5 Read textbooks, magazine articles and technical brochures to supplement your class lectures and notes. This information should then be added to your revision notes in a condensed form.
6 Answer the self-assessment questions found at the end of each chapter in this book and repeat them periodically; this will enable you to see which areas need listing for more concentrated revision.
7 Research fully topics you have listed for concentrated revision.
8 Periodically reread your notes and textbooks. This will help recall facts and reinforce them in your mind.
9 Mentally ask yourself questions on a particular topic. You can do this anywhere, even whilst travelling to and from work. Decide if your answer was suitable; if not, add the topic to your concentrated revision list.
10 Read past question papers and attempt timed answers. This is often done in class as a run-up to the examination.
11 In addition, use any technique you may have found successful in your previous studies.

Examination technique

As the examination day approaches, anxiety is normally the main problem to be encountered. Much of this can be overcome by the confidence gained from the knowledge that you have studied the course to the full and have undertaken a comprehensive revision programme. However,

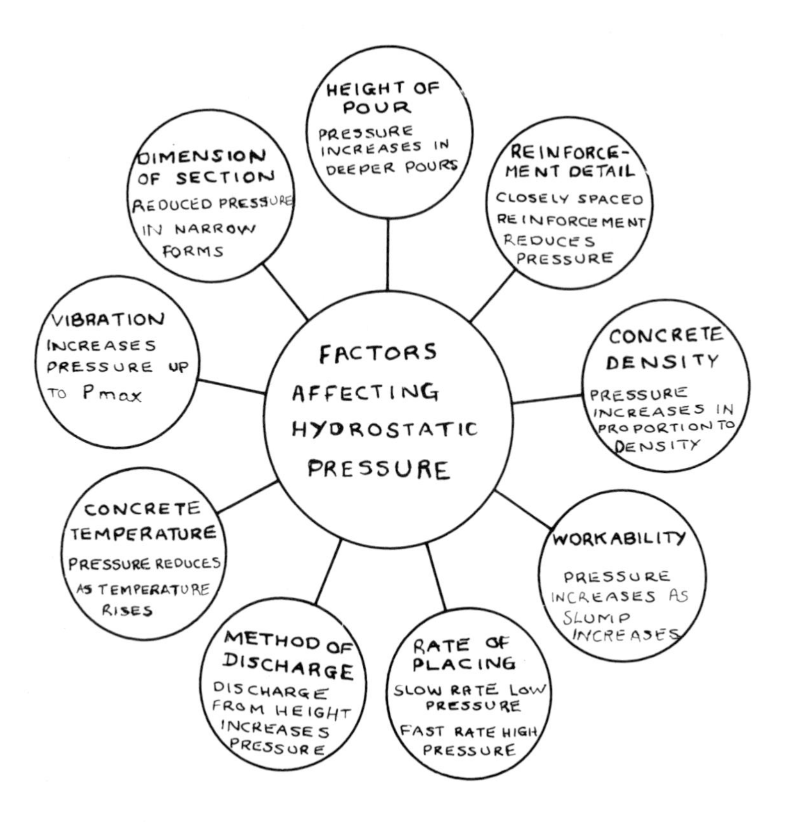

Figure 199 *Note network example*

do not be overconfident; a little anxiety is required to help you do your best. The use of the following checklist will aid you both before and during the examination itself.

Examination checklist

1. Arrive at the examination in plenty of time, ensuring that you have everything you are likely to need: examination card or number, pens, pencils, drawing equipment and electronic calculator.
2. Listen carefully to the invigilator's instructions.
3. Read the instructions at the top of your paper carefully.
4. Completely read through the paper. Many students at this stage underline key words in each question. In addition to the subject matter, look for words that indicate the length and precision of the expected answer:
   (a) Questions or parts of a question that start with 'name', 'list', 'suggest' or 'state' normally require a fairly brief answer of one or two words or a sentence at the most.
   (b) Words such as 'define', 'describe' or 'explain' require a longer answer, although these can also be shortened by the inclusion of 'briefly', for example 'briefly describe' etc.
   (c) 'Sketch' or 'find' call for less accurate answers than required from 'draw', 'develop' or 'calculate'.
   (d) Look out for questions that contain two indicating words, for example 'explain with the aid of sketches' etc.
5. Divide up the available time evenly between the number of questions to be answered, allowing say 12 minutes reading time at each end of the examination. For example, for a

three hour paper with six questions to be answered, a typical allocation of time would be:

(a) Reading time at the start of the examination: say 12 minutes.
(b) Reading and correction time at the end of the examination: say 12 minutes.
(c) Allocated time for each question: (180−24)/6 = 26 minutes.

6 Attempt your best answer first, as this will then give you more confidence to tackle the rest of the paper.
7 If you have been producing note network diagrams, these can be re-created at the top of your answer sheet to serve as a guide from which your answer can be developed. After completing your answer, the network can be crossed out.
8 Attempt the remaining questions in increasing order of difficulty. This will give you more time to think about and plan the harder questions subconsciously while you are completing the easier ones.
9 Keep to your time plan. When you reach the end of the time allocated to each question, stop writing even if you have not finished the answer. Leave a space before starting the next question so that if time permits you may come back and complete the answer later. It is far better to have attempted all the questions, even if some of the answers are incomplete, than to run out of time and leave the last two or so questions, which could happen if you spent five to ten minutes too long on each question.
10 When time is short your final answer can be put down in brief note form or even as a note network diagram; this will obtain you some marks at least.
11 Read through your paper at the end of the examination. This gives you a chance to spot and correct errors. Also you may have time at this stage to complete any unfinished answers.
12 Finally, ensure that you have put your name and student examination number on all your answer sheets and drawing paper.

The allocation of sufficient time to each question, and keeping to the time plan, are essential, since all questions which have to be answered should be answered. Do not fall into the trap of thinking that you have answered say the first four questions out of six fairly well, and so the remaining two can be left out as they do not count. It is very difficult to gain top marks for any question, no matter how well answered, but it is fairly easy to gain some, even for topics you do not think you know much about.

For example, in an examination of six questions carrying 10 marks each, someone who gained 9, 8, 7 and 6 marks respectively for the first four questions, but did not attempt the rest, would receive 50 per cent. However, if the final two questions had been attempted even in outline form, and received say 3 marks between them, then this would have increased the total to 55 per cent, which may make the difference between a pass and a credit grade.

You should now be ready to tackle your examinations. My best wishes for success go with you.

British Standards

The following British Standards are relevant to formwork and concrete practice:

CP 297: 1973	*Precast concrete cladding (non-load bearing)*
BS 12: 1978	*Ordinary and rapid hardening Portland cement*
BS 146:	*Portland blastfurnace cement* Part 2: 1973
BS 410: 1986	*Test sieves*
BS 812:	*Methods for sampling and testing of mineral aggregates and fillers* Parts 1–3: 1975 Part 4: 1976 Parts 101–102: 1984 Parts 103, 105, 106, 119: 1985
BS 877:	*Foamed or expanded blastfurnace slag lightweight aggregate for concrete* Part 2: 1973
BS 882: 1983	*Aggregates for natural sources of concrete*
BS 915:	*High Alumina cement* Part 2: 1972 (1983)
BS 1014: 1975	*Pigments for Portland cement and Portland cement products*
BS 1047: 1983	*Air cooled blast furnace slag coarse aggregate for concrete*
BS 1165: 1985	*Clinker aggregate for concrete*
BS 1199 and 1200: 1976	*Building sands from natural sources*
BS 1370: 1979	*Low heat Portland cement*
BS 1881:	*Testing concrete* Parts 101–122: 1983 Part 105: 1984 Part 115: 1986 Part 125: 1986 Part 201–203: 1986
BS 3148: 1980	*Methods of tests for water for making concrete*
BS 3681:	*Methods for sampling and testing of lightweight aggregates for concrete* Part 2: 1973 (1983)
BS 3797:	*Lightweight aggregates for concrete* Part 2: 1976 (1982)
BS 3809: 1971	*Wood wool permanent formwork*
BS 3892:	*Pulverised fuel ash* Part 1: 1982 Part 2: 1984
BS 4027: 1980	*Sulphate resisting Portland cement*
BS 4246	*Low heat Portland blastfurnace cement* Part 2: 1974
BS 4248: 1974	*Supersulphated cement*
BS 4340: 1968	*Glossary of formwork terms*
BS 4449: 1978 (1984)	*Hot rolled steel bars for*

	the reinforcement of concrete
BS 4461: 1978 (1984)	*Cold worked steel bars for the reinforcement of concrete*
BS 4466: 1981	*Bending dimensions and scheduling of reinforcement*
BS 4482: 1985	*Cold reduced steel wire for the reinforcement of concrete*
BS 4483: 1985	*Steel fabric for the reinforcement of concrete*
BS 4550:	*Methods of testing cement* Parts 0–1: 1978 Part 2: 1970 Parts 3–6: 1978
BS 5075	*Concrete admixtures* Parts 1–2: 1982 Part 3: 1985
BS 5328: 1981	*Methods for specifying concrete including ready-mixed concrete*
BS 5337: 1976	*Code of practice for structural use of concrete for retaining aqueous liquids*
BS 5507:	*Methods of test for falsework equipment* Part 1: 1977 Part 3: 1982
BS 5950: 1982	*The structural use of steelwork in building* Part 1: 1985 Part 2: 1985 Part 4: 1982
BS 6100:	*Glossary of building and civil engineering terms* Part 0–2: 1984 Part 3: 1986 Part 4: 1984
BS 6089:	*Guide to assessment of concrete strength in existing structures*
BS 8110:	*Structural use of concrete* Parts 1–3: 1985

Index

63442